Guide to Good Practice in
The Management of Time
in Major Projects

Dynamic Time Modelling

Second Edition

The Chartered Institute of Building

WILEY Blackwell

Registered Office(s)
John Wiley & Sons, Inc., 111 River Street, Hoboken, NJ 07030, USA
John Wiley & Sons Ltd, The Atrium, Southern Gate, Chichester, West Sussex, PO19 8SQ, UK

Editorial Office
The Atrium, Southern Gate, Chichester, West Sussex, PO19 8SQ, UK
For details of our global editorial offices, customer services, and more information about Wiley products visit us at www.wiley.com.

Wiley also publishes its books in a variety of electronic formats and by print-on-demand. Some content that appears in standard print versions of this book may not be available in other formats.

Library of Congress Cataloging-in-Publication Data
Names: The Chartered Institute of Building (Great Britain)
Title: Guide to good practice in the management of time in major projects :
  dynamic time modelling / by The Chartered Institute of Building.
Description: Second edition. | Hoboken, NJ, USA : Wiley, 2018. | Includes
  bibliographical references and index. |
Identifiers: LCCN 2017052026 (print) | LCCN 2017055842 (ebook) | ISBN
  9781119428459 (pdf) | ISBN 9781119428435 (epub) | ISBN 9781119428398 (pbk.)
Subjects: LCSH: Building–Superintendence. | Production scheduling.
Classification: LCC TH438.4 (ebook) | LCC TH438.4 .G85 2018 (print) | DDC
  690.068/5–dc23
LC record available at https://lccn.loc.gov/2017052026

Cover design by Wiley
Cover image: The CIOB wish to thank Freeform Advanced 4D Modelling for the images used in the design of the front cover.

Set in 10/13pt ITCFranklinGothic by SPi Global, Chennai, India

10 9 8 7 6 5 4 3 2 1

Contents

Contents

Preface

As with a well-cut diamond, project success has many facets that work in unison to create a brilliant result; and as with a well-cut diamond, some facets contribute far more to the overall impression than others. This book is focused on a major facet of project success, effective project time management, which underpins the cost-effective use of resources and contributes to achieving a time- and cost-efficient project without compromising quality, safety or risk management objectives.

This book is the second edition (retitled to better reflect its objective) of CIOB's highly successful *Guide to Good Practice in the Management of Time in Complex Projects,*[1] first published in 2011. Since 2011, leading authorities globally have increasingly recognised 'schedule is king'[2] and have placed increased emphasis on the proactive management of time, using dynamic modelling, as a precursor to project success. In keeping with this theme, the USA Government Accountability Office (GAO) published its *Schedule Assessment Guide: Best Practices for Project Schedules*[3] in 2015 to complement its well-regarded *Cost Estimating and Assessment Guide.*[4] The CIOB has also updated its *Complex Projects Contract, 2013*[5] as the *Time and Cost Management Contract suite 2015,*[6] including back-to-back Consultancy Appointment and Subcontract, and many other standards and guides requiring effective project time management have also been updated or published since the first edition of the Guide was released.

This *Guide to Good Practice in the Management of Time in Major Projects – Dynamic Time Modelling* does not seek to duplicate these standards; rather it provides the practical and rigorous framework needed to guide scheduling practice to achieve the objectives defined by these standards. Applying the guidance contained in this book will offer any project team the best way to achieve the effective management of the time available to complete their project, conform to recognised good practices, and consequently create the best opportunity for a successful project outcome.

---

[1] http://www.ciob.org/time-management (accessed 10 December 2017).
[2] Stephen Gumley, former CEO, Defence Material Organisation.
[3] https://www.gao.gov/products/GAO-16-89G (accessed 10 December 2017).
[4] http://www.gao.gov/products/GAO-09-3SP (accessed 10 December 2017).
[5] http://www.ciob.org/insight/contract-complex-projects (accessed 10 December 2017).
[6] http://www.ciob.org/insight/time-and-cost-management-contract-suite (accessed 10 December 2017).

Introduction to Second Edition

The *Guide to Time Management in Major Projects – Dynamic Time Modelling* (the Guide) is a revised edition of what was previously the CIOB's *Guide to Good Practice in the Management of Time in Complex Projects*, published in 2011.[1] The name has been changed to reflect more clearly the core strengths of the Guide and its application to the management of major projects; however, whilst the Guide is focused on construction and engineering projects, the concepts and procedures can be adapted for use in any type of project.

The Guide is a practical treatise on the processes to be followed and standards to be achieved in the effective management of time. Subject to the amendment of existing forms of contract to remove inconsistencies, the Guide can be used in any jurisdiction, under any form of contract, with any type of project.

The Guide promotes competence in critical path network modelling, resource allocation and productivity analysis by the use of a dynamic time model (DTM).[2] The Guide does not recommend a single density,[3] static baseline target programme, but requires a dynamic critical path network, in differing densities, updated and revised on the rolling wave principle which constantly predicts the currently attainable completion date, sectional completion dates and key dates as a result of the current sequencing.

Although the Guide is not based upon any contractual regime or procurement process, the CIOB's 'Time and Cost Management Contract' suite (2015)[4] has been written for use with the Guide to provide a uniform approach to cost and time risk management from initiation to completion of major building and engineering projects in accordance with the Guide's recommendations.

---

[1] http://www.ciob.org/time-management (accessed 10 December 2017).

[2] A dynamic time model is a critical path network of which the essential characteristics are: 1.a high-quality critical path network without any constraints that will inhibit the schedule from reacting dynamically to change; 2.a combination of the short-term look ahead using resource and location based logic in High Density with the schedule for the whole of the work using activity based logic in Medium Density and Low Density; 3. activity durations in the High Density part of the network are calculated by reference to the planned resources and their expected productivity and updated with records of the resources actually used and the amount of work actually achieved in the update period; 4. the schedule is revised to incorporate fuller and better information as it become available on a 'rolling wave' principle going forward so that the next 3 months work is always scheduled in High Density; 5. the schedule is impacted by intervening events that have occurred, are occurring or are likely to occur in the future in order that their predicted effect can be managed. See also Parts 1 and 3 and Appendix 4 for a description of the dynamic time model in use.

[3] Density refers to the level of detail contained in the schedule, see Section 4.2.

[4] See http://www.ciob.org/insight/time-and-cost-management-contract-suite (accessed 10 December 2017).

Acknowledgements

The Guide is the result of the combined efforts of the editorial committee, comprising:

■ Robert Clark, BSc (Hons) MRICS FCIOB, Director, Failand Consultancy Ltd. robac49@gmail.com

■ Trevor Drury, Barrister, MBA, PG Dip Project Management, FRICS, FCIOB, MCIArb, 12 Old Square Chambers, and Managing Director, Morecraft Drury. trevor.drury@morecraft-drury.com

■ Dr David-John Gibbs, MEng, EngD, ICIOB, Managing Consultant for BIM, HKA. davidjohngibbs@hka.com

■ Paul Kidston, FCIOB, Project Control Director, Costain Group. paul.kidston@costain.com

■ Keith Pickavance, LLb (Hons.) Dip Arch, Dip IC Arb, PPCIOB, keithpickavance@live.co.uk (editor)

■ David Tyerman, MBA, LLM, FCIOB, Director, Kingsfield Consulting. david.tyerman@kingsfieldconsulting.com

■ Patrick Weaver, FCIOB, Managing Director, Mosaic Project Services Pty Ltd. patw@mosaicprojects.com.au

Table of Figures

1 Introduction

1.1 Core principles of time management

1.1.1 Effective time management is essential to resource management, cost management and the allocation of liability for slippage, its recovery and accountability.

1.1.2 In order to achieve effective time management there must be:

- a carefully considered planning method statement that directs and maintains the use of the dynamic time model (DTM);[1]

- a competent appraisal of the risks which are likely to have a significant effect on the progress of the work in the future;[2]

- a design which permits the work sequences that are likely to be severely disrupted and/or delayed by foreseeable risks to be separated into parallel, rather than sequential paths;[3]

- a dynamic time model for the project against which progress, or lack of it, can be measured, accounted for and the consequences of it predicted;[4]

- a practically achievable strategy for dealing with intervening events during the design, procurement and construction processes.[5]

1.1.3 The word 'programme', often used in the past to describe a static baseline target (usually printed on paper and containing a list of dates on which the proposed activities might be carried out) is not used in connection with the management of time in major projects.

1.1.4 The word 'schedule' is used in the Guide to describe a dynamic time model comprising the computerised calculated activity dates and logic. The process is to be referred to as scheduling and the occupation that of the scheduler. The schedule is manifest in an editable computer file.

1.1.5 Planning and scheduling are separate disciplines. Project planning is largely an experience-based art, a group process requiring contribution from all affected parties for its success. The output from planning is documented decisions on how the work of the project will be accomplished. Scheduling is the science of using mathematical calculations and logic to model the project plan to predict when and where

---

[1] See Section 1.2, Part 3 and Appendix 4 for a description of the DTM in use.
[2] See Appendix 1 for a listing of those generic risks normally borne by the employer.
[3] See Appendix 2 for case studies illustrating the importance of considering planning strategy during the design stage.
[4] See Section 1.2, Part 3 and Appendix 4 for a description of the process and use of the DTM.
[5] See 'Managing the dynamic time model' at Part 5.

Guide to Good Practice in the Management of Time in Major Projects: Dynamic Time Modelling, Second Edition. The Chartered Institute of Building.
© 2018 John Wiley & Sons Ltd. Published 2018 by John Wiley & Sons Ltd.

work is to be carried out in an efficient and time effective sequence. The output from scheduling is a dynamic project schedule.

1.1.6 Planning must precede scheduling. They cannot be carried out in parallel, nor can scheduling precede planning.

1.1.7 Schedule preparation must be a quality assured process against a standard which will ensure the integrity of the schedule, so that it can function as a time model.

1.1.8 The schedule (and any revisions and updates) should be audited, independently of any contractual obligation, for integrity and technical competence.

1.1.9 Time management starts with the conceptual design of the project. If the design does not permit the work to be carried out efficiently, and hence is not time effective, no procurement strategy will rescue it.[6]

1.1.10 Time management of major projects necessarily encompasses the management of design, manufacture, procurement, subcontract and separate contractor work packages, information flow, quality control, safety management and the achievement of multiple key dates, sectional completion dates and multiple projects.

1.1.11 A time risk appraisal is to be carried out at inception and constantly updated throughout the life of the project.

1.1.12 Time contingencies for the employer's, the design team's and the contractor's risks must be a part of the strategy for effective time control. The basis for assessing each of the contingencies must be described.

1.1.13 In order to identify the cost advantages of various development strategies, and hence the most efficient overall approach to the work, the value of the time contingencies allowed against foreseeable risks must be included in the project budget.

1.1.14 The Guide differentiates between the development schedule, prepared before a contractor is appointed, and the working schedule used in connection with construction.

1.1.15 The development schedule cannot be prepared in one process at a single density, or degree of detail, at inception. It must be prepared in varying densities[7] consistent with the information available from time to time, and reviewed and revised at regular intervals, as better and more certain information becomes available.

1.1.16 The working schedule must follow from the development schedule and must also be prepared in varying densities consistent with the information available from time to time. It must also be reviewed and revised at regular intervals as better and more certain information becomes available.

1.1.17 Consultants', specialist contractors' and subcontractors' schedules must be prepared in the same software as the development and working schedules and integrated with them.[8]

1.1.18 The work to be carried out in the short term must be scheduled according to the resources actually available and the productivity expected from the resources for the various work types to be carried out.

---

[6] For examples of this principle in practice, see Appendix 2 'Case studies in strategic planning'.
[7] For guidance on schedule density, see Section 4.2.
[8] For guidance on schedule integration, see Section 4.8.

1.1.19   Where the work is to be or has been designed in a BIM environment, the 3D virtual model can be used to assist with quantifying the work for each activity in the High Density part of the schedule.

1.1.20   If the 3D virtual model is linked to time-related information to produce a 4D model, the alignment of objects in the 3D model to the schedule can help to reduce difficulties arising out of incompleteness of design and/or schedule information.

1.1.21   Baselines exist to permit status reporting rather than the dynamic management of time. Therefore, where baselines are used they must be maintained to reflect changes to the content of the work and development of schedule density as the project proceeds.

1.1.22   Progress monitoring techniques, which are rooted in comparison of estimated progress against a static baseline target, have limited value in the management of time in major projects (in which the work content, resources and sequence necessarily change from time to time).

1.1.23   Progress records must be kept on a database that will provide instantaneous access and retrieval of as-built data for the purpose of checking the reliability of productivity assessments in varying repetitive work cycles, and hence the reliability of Medium Density and Low Density scheduling.

1.1.24   Because progress data will be recorded only against the fully resourced, High Density part of the schedule, the as-built record will provide data standards and productivity feedback for future benchmarking, which will improve predictability and hence reliability of short-term scheduling.

1.1.25   Quality control and information flow should be managed via the same database as that used for the maintenance of progress records.

1.1.26   The dynamic management of time necessarily includes the management of the consequences of delays to progress caused by of all types of events, referred to as 'intervening events' in the Guide.

1.1.27   Activities representing the intervening events are to be inserted into the relevant schedule and the likely consequences of the event calculated. This assessment should be undertaken as soon as the intervening event is identified, with adjustments made as the situation unfolds. Where practical, mitigation should be planned, agreed and included in the schedule.

1.1.28   The Guide contains no guidance for the approximation of a 'fair and reasonable' extension of time or compensation for delay-related costs that might be incurred.

1.1.29   The time management strategy is to be set down in writing in a regularly updated method statement, which is to deal with, amongst other things, the stated strategy and assumptions adopted for:

- project planning

- risk management

- schedule preparation

- schedule review and revision

- progress update

- record keeping

- quality control

- communications.

Part 1 - Introduction

1.2 The dynamic time model

1.2.1 Production and use of the DTM differs significantly from the traditional method of programming or scheduling. The following list outlines some of the differences.

	Traditional programme or schedule	The dynamic time model
Design risk management	A critical path network schedule is not used to manage the design process and, as a consequence, the design programme or schedule often has faulty or incomplete logic, if any at all. As a result, it is not possible to use the design programme or schedule to manage change, predict consequences or manage risk during the design process.	The time management process commences with the development schedule constructed as a DTM. The development schedule must be a high quality, fully networked schedule of the design and other preconstruction processes, that can be used to predict consequences and manage change during the design and pre-contract services, and which subsequently may be used to inform, in whole or in part, the contractor's working schedule.
Construction risk management	The programme or schedule used by most contractors to manage their work is not required to be a high quality critical path network and, as a consequence, those produced often have faulty or incomplete logic, if any at all. As a result, it has not been possible to use the contractually required programme or schedule to manage change, predict consequences or manage risk during the construction process.	The DTM must be a high quality, fully networked schedule that can be used to predict consequences and to manage change.
Transparency and communication	The contractually required programme or schedule is not always the product of database-driven scheduling software and is generally issued on paper or PDF, without the employer and its consultants having access to the assumptions, data or calculations used to produce it.	The DTM is required to be the product of database-driven scheduling software with a planning method statement identifying the assumptions and calculations in it, and to which full electronic access may be given to all stakeholders and its consultants.
Programming or scheduling for incomplete design	The contractually required programme or schedule describes the whole of the works at a uniform density often in thousands of lines of data. Typically, a high level of detail is required by most construction and engineering contracts, either without provision for (or forbidding) revision without approval.	Work that has not yet been designed cannot be programmed or scheduled in anything other than outline. Accordingly, the DTM comprises activities in different densities, the detail of which depends upon the degree to which the design, resources and productivity can be ascertained. The DTM is then to be revised on a 'rolling wave' basis as more detailed information becomes available

	Traditional programme or schedule	The dynamic time model
Transparency of the short-term look-ahead	The contractually required programme or schedule is not used by the contractor for the purposes of managing the work on site. The contractually required programme or schedule is not always produced before the work has started, changes to it are discouraged and it is used, primarily, in disputes with the client. The contractor may produce an overall programme or schedule for compliance with the contract, but detailed programming or scheduling of the work on site, if done at all, tends to be by use of a separate short-term look-ahead, which is not embedded in the contractually required programme or schedule, is not produced in the same software and access to it is not available to the employer and/or its consultants.	The DTM starts in the design process and continues, through construction to completion. It incorporates a transparent short-term look-ahead in the High Density part of the working schedule for managing the work on site, in which the activities are calculated by reference to the planned resources and productivity, and to which the employer and its consultants may have electronic access.
Integrity of the short-term look-ahead	The contractor's short-term look-ahead is not constructed as a network, is not resource loaded, does not contain any productivity data, and the activity durations are estimated against unspecified data.	The High Density part of the DTM must be resource loaded and the activity durations must be calculated by reference to the resources planned to be applied and the productivity they can be expected to achieve.
Revision of the programme or schedule	The contractually required programme or schedule is not required to be revised, and on major projects, it tends to quickly become out of date and fall into disuse.	The DTM requires that the critical path network be regularly revised to take into consideration the latest available information, on a 'rolling wave' basis, to keep it up to date.
Recording progress of work	Progress achieved is either estimated subjectively, gauged against the performance to intermittent milestones, or calculated by reference to the value of the work estimated to be fit for payment.	The DTM (which has a fully resourced network for the work to be carried out in the short term) requires, as its progress records, databased details of the resources actually used, the quantity of work actually performed and the time period of the record.
Updating the programme or schedule with progress achieved	If progress is marked up on the programme or schedule at all, it is often marked up on a 'jagged line' principle that does not identify the effect of progress achieved on future work.	The DTM requires that the critical path of the updated schedule is recalculated to identify the effect of progress achieved on the key dates, sectional completion dates and the completion date, and to identify any corrective sequence that must be followed.

	Traditional programme or schedule	The dynamic time model
Allowing time contingencies for unknowns	Although it is common for prime cost, provisional cost and cost contingency sums to be included in project budgets, historically no allowance for spending such sums has been required to be included in the contractually required programme or schedule.	The DTM requires that appropriate time contingencies be included for all undefined work that is expected to be needed, including that arising out of perceived risks.
Extensions of time and compensation	Because the actual progress of the work frequently differs significantly from the planned sequence in the contractually required programme or schedule and it is not revised to identify additional or changed work, progress, if recorded at all, is rarely recorded in a way that relates to the contractually required programme or schedule. As a consequence, claims for extension of time and compensation are often framed subjectively and left until after the opportunity for risk management has passed.	The DTM requires progress to be calculated, and revisions to be made regularly to keep the schedule up to date and for the effects of change to be calculated objectively by incorporating the intervening event and its effects into the critical path network. This creates contemporaneous information upon which informed and calculated decisions can be made, with all stakeholders and their consultants.
Benchmarking	The contractually required programme or schedule bears such little resemblance to the way the work is actually carried out that it has no use at all after completion, save to be resurrected by one side or the other in dispute proceedings.	When the works are completed, the DTM contains all the resource and productivity data of the work actually constructed and can be used to good advantage in benchmarking, to improve scheduling and cost estimating on other projects.

1.3   Mission statement

1.3.1   The primary purpose of the Guide is to set down the standards necessary to facilitate the effective and competent management of time in construction projects.

1.3.2   The Guide requires a planning strategy that recognizes the predicted out-turn time and out-turn cost and not just contract time and cost.

1.3.3   The Guide defines the standards by which project schedules will be prepared, quality controlled, updated, reviewed and revised in practice.

1.3.4   The Guide describes the standards of performance that should reasonably be required of a project scheduler. It will also form the basis for the education of project schedulers.

1.3.5   Without compromising its primary purpose, the Guide can be developed as a strategic planning and scheduling reference document capable of wide application.

1.4 Genesis of the Guide

1.4.1 The continuous pursuit of excellence in the management of construction and engineering projects is the key to more effective collaboration, the continued satisfaction of the industry's client requirements and the sustained delivery of successful projects in the 21st century.

1.4.2 With a view to examining the state of the industry's practices in time management, between December 2007 and January 2008 the CIOB conducted a survey of the industry's knowledge and experience of different methods of project control and time management, record keeping, monitoring and training.[9] The results indicated a wide disparity between the experience of the respondents and good practice in time management.

1.4.3 In the light of the results of that research, and with a view to reducing the incidence of delayed projects, the CIOB has initiated the Guide as a part of the initiative in encouraging excellence in the management of construction, increasing awareness of the importance of project planning and scheduling in the industry as a whole and and, in particular, with regard to the management of time in major projects.

1.5 Purpose of the Guide

1.5.1 The growth in training, education and skill levels of the industry in the use of time management techniques has not kept pace with the technology available. There is, however, a trend towards developing more complicated projects with contracts that are increasingly punishing, if not executed efficiently, with effective time management and project controls.

1.5.2 It is apparent that, since the 1980s, the construction industry has experienced:

■ Design and Build, Guaranteed Maximum Price, and Engineer, Procure and Construct Contracts and other methods of procurement which require the contractor to take more risk than in traditional forms of contracting;

■ specially incorporated companies as employers for particular projects which have limited access to additional funds and are intended to be liquidated once their purpose is fulfilled, and;

■ the requirement to deliver more technologically sophisticated solutions in shorter timescales and within tighter financial constraints.

1.5.3 However, this is not a guide to project risk, value, or other management specialities. The purpose of the Guide is to set down the strategy and the standards necessary in order to facilitate the effective and competent management of time in major projects.

1.5.4 Time modelling with the use of computers to develop a framework by which the consequences of change and other intervening occurrences can be managed technically and objectively has been available since the early 1960s. However, it was only in the last few years of the 20th century that the necessary computing power and software became commonly available to facilitate the objective measurements of project deliverables.

---

[9] *Managing the Risk of Delayed Completion in the 21st Century* (2008), Chartered Institute of Building (available from http://www.ciob.org/time-management (accessed 9 December 2017)).

1.5.5   Developments in hardware, software and communications services in the last decade of the 20th century have rendered it virtually impossible in the 21st century to conduct any business efficiently without the use of computers and electronic services.

1.5.6   When the first edition of the Guide was written, in 2011, it was apparent that the construction industry used those resources intensively in design, in manufacture, in procurement, in assembly, in finance and in virtually every field other than the management of time. The CIOB's research showed that time management was generally pursued intuitively, and programmes, if used at all, were used only as a static baseline target against which a failure to succeed could be gauged and reported upon.

1.5.7   Whilst it is apparent that simple projects can be managed intuitively by experienced construction managers against a static baseline target programme, it is also apparent that the management of major projects cannot be successfully managed intuitively. Attempts to manage time on major projects by reference to estimated progress against a static baseline target programme will result in failure because there are simply too many consequential possibilities in major projects for time to be managed intuitively. A more scientific approach using a dynamic time model and calculation is required to assess the consequences of express and implied changes in the work, and the effect of other intervening events on the multiplicity of activities in a continually changing time frame.

1.5.8   Just because the Guide focuses on the requirements of major projects, this does not mean that what is recommended here cannot be adopted for simple projects if that is what the employer and/or contractor or consultants wish. However, it does mean that managing time by comparison of estimated progress against a static baseline target programme is simply not good enough for major projects.

1.6   Risk management

1.6.1   Proactive time risk management requires the employer to engage actively in the strategic design of the overall project and to employ an approach to obtaining the services of competent consultants and contractors that maximises the probability of achieving completion of the works on time. The key areas requiring employer engagement are:

■   adoption of a project strategy that optimises the balance between time and cost risks. Appendix 2 to the Guide offers a number of case studies focused on project strategy;

■   adoption of a form of contract that encourages the proactive management of time and allows the concepts outlined in the Guide to be used effectively;[10]

■   balancing the allocation of risk between the parties to promote the best opportunity for effective time risk management (see Section 2.3);

■   including time risk management in the selection criteria for selection of consultants and contractor to design and construct the works (see Section 2.2).

---

[10] The principles of the Guide are the foundation for the CIOB's Time and Cost Management Contract 2015, which together with the Consultancy Appointment and Subcontract constitute the suite of Time and Cost Management Contracts and associated documents. It is available free from http://www.ciob.org/insight/time-and-cost-management-contract-suite (accessed 10 December 2017).

1.7 Planning and scheduling

1.7.1 Project planning and scheduling are allied disciplines, but they are not one and the same.

1.7.2 Project planning is a team operation, involving the construction management team, cost control team, design team and project planner in creating the project development and delivery strategy. There are fundamental aspects of planning that require a conceptual approach similar to that of design. It requires experience, vocabulary, communication and imagination to craft the overall approach and strategy for project design, development and delivery.

1.7.3 Project planning involves decisions concerning:

■ the overall strategy of how the work process is to be broken down for control;

■ how the control is to be managed;

■ what methods are to be used for design, procurement and construction;

■ how commissioning, testing and transfer to the client will be accomplished;

■ the strategy for subcontracting and procurement;

■ the interface between the various participants;

■ the zones of operation and their interface;

■ maximising efficiency of the project strategy with respect to cost and time; and

■ threat and opportunity risk management.

1.7.4 Scheduling is a mixture of art and science, involving the interpretation of the results of project planning to ascertain, amongst other things, the activities required to accomplish the work, their logical sequence, and start and finish dates. Scheduling is usually performed by the use of software that facilitates the fast and efficient manipulation of the project planning data for the purposes of time and cost risk management. In effect, the schedule is the construction manager's time allocation tool, the employer's risk management tool and the contract administrator's calculator.

1.7.5 Project scheduling is the art and science of putting the decisions made at the project planning stage into a database in order to:

■ enable the scheduler to allocate contract calendar periods and resources to the various sections of the work;

■ determine a logical sequence for accomplishing the work;

■ allocate time contingency periods;

■ calculate the overall project duration, float and resource requirements; and

■ present the strategy in a form acceptable to the contractor, employer and the contract administrator as a process management tool.

1.7.6 In the process of converting the plan into a schedule (within a framework, which will react dynamically to change, so as to facilitate the management of time throughout the life of the project), the scheduler should determine:

■ the duration of each activity;

■ the party who will perform each activity;

■ the resources to be applied to each activity; and

■ the method of sequencing of one or more activities in relation to other activities.

1.7.7   It is not good practice to plan the work whilst attempting to schedule it. In the same way that it is possible to start designing a building at the same time as preparing the working drawings and other production information, it is equally possible to perform the project planning operation whilst scheduling. However, in neither case is such an approach likely to produce, on the one hand, a satisfactory design and consistent production information, nor, on the other, a satisfactory project planning solution and effective schedule.

1.7.8   Accordingly, the Guide recommends that the project planning function is performed first and the scheduling operation carried out in accordance with the established strategic project plan and a planning method statement that explains the assumptions and calculations used in creating the schedule.

1.7.9   In essence, the prior planning procedure should encompass:

■   familiarisation

■   outline plan

■   strategic plan

■   detailed plan and planning method statement

■   risk assessment

1.7.10   Only when that has been completed should the project scheduling process commence.

1.7.11   The importance of the project schedule to time management cannot be overemphasised. Without a time model that will react dynamically to change, it is not possible, except intuitively, to forecast when work is to be carried out, nor is it possible to assess its criticality, nor the impact of progress achieved and intervening events on successor activities, resources required, key dates, sectional completion and project completion.

1.8   The planning method statement

1.8.1   Consultation and effective communication are prerequisites to creating a meaningful and effective schedule based on an agreed planning method statement. Whereas the purpose of the schedule is to describe how the work of the project is to be accomplished, the purpose of the planning method statement is to facilitate understanding of the schedule and cooperation between the participants.

1.8.2   The planning method statement is the record of the assumptions made and the calculations used in transferring the planning strategy into the schedule at its data date. It should make clear what constraints have been identified, what assumptions have been made in the process of risk management, planning, scheduling, revision and update of the schedule, and the reasoning underpinning those constraints and choices.

1.9   The project scheduler

1.9.1   The job of the project scheduler is to devise and maintain the schedule model from overview to micro-level and to manage practical and effective time control from commencement to completion of the project.

1.9.2   In order to verify the accuracy of updated schedules, the scheduler must thus be able to advise on and manage the development of the overall schedule, the making and retrieval of progress records and, in order to keep the project and construction management team informed, the project scheduler must be able to prepare an as-built schedule of work carried out, progress schedules and progress-related data, for the purpose of progress reporting during the course of the works, and predictive schedules to test or demonstrate future courses of action.

1.9.3   Apart from creating the schedule at inception, the project scheduler will be engaged in editing, reviewing, revising and updating the schedule throughout the life of the project. Reviewing and revising schedules in the light of better information also brings the need to be able to advise on and manage the writing, revision and editing of the planning method statement.

1.9.4   When change is imposed, the project scheduler must be able to identify contemporaneously the effect of the change, and where delay or disruption to the planned sequence of work occurs, to advise the project planner and other members of the project management team on the likely effect of possible recovery strategies.

1.10   Time management

1.10.1   Time management is the science of identifying, from time to time, what, in the light of current status and information, the completion of a sequence, key date, sectional completion date or overall completion date is likely to be. If achievement by that date is not what is required, then time management requires that in the light of the information then available, the planning strategy and relevant schedule for the future conduct of the work are amended so as to plan to achieve the dates required.

1.10.2   Accordingly, the schedule is to be used for identifying the following intentions:

- the duration and sequence of the activities required to accomplish the work and the interface with any other contracts incidental to the work;

- the resources required to accomplish the work as planned;

- dates and logic by which the information described in the information release schedule, information request schedule, or any other request for information, is to be supplied in relation to the work dependent upon such information;

- dates and logic by which plant, materials or goods are to be supplied, or work to be carried out by the employer, or those engaged or employed by them in relation to the work dependent upon them;

- any time contingency required by the contractor, any subcontractor and/or supplier in relation to any activity, sequence of activities, or key dates, or the contract requirements for any sectional completion dates and the completion date;

- any time contingency required by the employer, or any directly employed contractor, or consultant, in relation to any activity, sequence of activities, key dates, the contract requirements for any sectional completion dates and the completion date;

- free float and total float that are available to be used by the contractor and/or the employer for managing the resequencing of the work or redeployment of resources from time to time;

- the actual resources used on all activities carried out to date;

■ the degree of progress actually achieved on all activities to date;

■ the likely and actual effect of any delay to progress, if any, on the completion of any sequence, and the contract requirements for any key dates, sectional completion dates and/or the completion date, caused by the rate of progress achieved or an intervening event;

■ the likely effect of any proposed acceleration, or recovery measures on any such sequence, the contract requirements for any key dates, sectional completion dates and/or the completion date.

1.11 Building information modelling

1.11.1   Building information modelling (BIM) is applicable not only to building works but also other construction and engineering works of all types, above and below ground and subsea.

1.11.2   The contractual requirements for the preparation of an information model may vary in detail from project to project, but in essence they encompass a virtual 3D model comprising parametric objects.

1.11.3   Parametric objects are more advanced than computer-aided design (CAD) based lines. Instead of using the 2D plan, section and elevational approach as representations of 3D projects, objects placed in the virtual 3D model recognise size and space, and may include other information related to cost, facilities management, embedded carbon and many other facets of the object. A parametric object can be created for all components of the project, offering significant benefits in the processes of design coordination and change management.

1.11.4   Parametric objects can be linked to or embedded with data which, depending upon the extent to which the data is developed, may include:

■ specification

■ materials

■ resources

■ productivity

■ value

1.11.5   Parametric objects that are linked to a schedule activity ID (which may also include resources and productivity data) are referred to as 4D models. Those that also include value data are known as 5D models.

1.11.6   When adopting 4D modelling, care should be taken to ensure interoperability and transparency between the 3D modelling software and the scheduling software used on the project. Care must be taken in establishing a two-way approach to data flow, and ensuring that workflow protocols are established to create and maintain transparency.

1.11.7   The use of BIM systems requires the use of integrated IT systems in a much wider context from inception, through design, procurement and delivery.

2 Strategy

2.1 Planning method statement strategy

2.1.1 Consultation and effective communication are prerequisites to creating a meaningful and effective schedule based on an agreed planning method statement. Whereas the purpose of the schedule is to describe how the work of the project is to be accomplished, the purpose of the planning method statement is to facilitate understanding of the schedule and cooperation between the participants.

2.1.2 The planning method statement is the record of the assumptions made and the calculations used in transferring the planning strategy into the schedule at its data date. It should make clear what constraints have been identified, what assumptions have been made in the process of risk management, planning, scheduling, revision and update of the schedule, and the reasoning underpinning those constraints and choices.

2.1.3 At inception the scheduler will draft an appropriate planning method statement with the input and cooperation of others. Those who may participate will depend upon the nature of the project, but might reasonably include:

- the employer
- the design team
- the risk manager
- the project manager
- the construction manager
- the project scheduler
- the health and safety manager
- the planning manager

As the planning method statement evolves, input of others will be required, for example:

- other directly employed contractors
- utilities, statutory undertakings and third party projects
- specialist designing subcontractors

Guide to Good Practice in the Management of Time in Major Projects: Dynamic Time Modelling,
Second Edition. The Chartered Institute of Building.
© 2018 John Wiley & Sons Ltd. Published 2018 by John Wiley & Sons Ltd.

■ trade package contractors

■ domestic subcontractors

■ specialist suppliers

■ specialist design consultants

2.1.4 In developing the planning method statement, consideration must be given to both the size and complexity of the work to be managed. Some aspects of the work that could indicate a degree of complexity are outlined in Appendix 3.

2.1.5 Because it will have a life independent of those who, from time to time, may be required to work upon it, it is important that the planning method statement is designed for ease of use by everyone, including those who have little involvement in the work of the project.

2.1.6 The content of the planning method statement will change during the development of the project and must be designed to accommodate and track change in subject matter content and source, without compromising transparency between the planning method statement and other time-related information.

2.1.7 Any amendments made to the underlying assumptions contained in the planning method statement will also need to be carefully recorded in a clear and concise manner.

2.2 Consultant and contractor selection strategy

2.2.1 The employer sets the framework for managing time-related risks, but in large part the implementation of time risk management relies on the capabilities of the employer's consultants and contractors. Accordingly, it is prudent to seek to engage consultants and contractors with demonstrated time risk management abilities.

2.2.2 The information (to be provided at the time of tender) to permit a selection process that balances overall cost with time management capability should include:

■ The consultant's and/or contractor's strategy for accomplishing their work in a timely manner.

■ The anticipated resources and estimated productivity required to achieve the consultant's and/or contractor's work as scheduled in the first three months of the project.

■ The contingencies allowed for the consultant's and/or contractor's identified risks (including those risks specifically allocated under the relevant contract).

■ The strategy for implementation of effective risk reduction.

2.2.3 It is generally the case that even after demonstrating adequate resources and the inclusion of appropriate contingencies, the probability of completion of the relevant contract within the permitted contract period will still be less than 100%.

2.2.4 By requiring the consultants and contractors to demonstrate the probabilistic outcomes of their tender and the risk they are accepting (which should be compensated by the profit they are expecting) will provide assurance to both parties that the contracting party is aware of its risks and capable of absorbing the adverse risk to which it is exposed.

2.3 Contracting strategy

2.3.1   Subject to the amendment of existing forms of contract to remove inconsistencies, the Guide can be used in any jurisdiction, under any form of contract and with any type of project. Whilst the Guide is focused on construction and engineering projects, the concepts and procedures in the Guide can be adapted for the management of time in any industry.

2.3.2   Although the Guide is not based upon any contractual regime or procurement process, the CIOB's 'Time and Cost Management Contract' suite (2015)[1] has been written for use with the Guide to provide a uniform approach to time and cost risk management from initiation to completion of major building and engineering projects in accordance with the Guide's recommendations.

2.3.3   An effective contracting strategy will dictate an appropriate balance of risk and cost. An appropriate balance of risk and cost requires the employer to transfer to the contractor those risks which the contractor can most effectively manage and retain for itself those risks which the contractor cannot manage.

2.3.4   The contract for the project is the mechanism for the apportionment of risks between the parties, and consequently requires careful drafting. The appropriate balance of risk and cost depends on the circumstances of the project, but there are three principles the employer should always consider:

- It is impossible to transfer all risk to the contractor; ultimately, the employer always bears the risk of late delivery.

- Competent contractors will charge a commercially realistic premium for each risk they are required to carry.

- Successful contractors are successful because they only accept risks they can manage; if they perceive a contract as being 'too risky', their tender for the work will not be competitive.

2.3.5   The terms of the relevant contract give effect to the allocation of risk between the parties to it:

- In determining the nature of the risk being managed, it is important that attention be paid to the precise phraseology used.

- Risks peculiar to a project and their apportionment should be set down in the contract conditions.

- Care is required to ensure that the adopted form of contract implements the balance of risk required by the employer. Under some forms of contract, a risk that is ordinarily the employer's passes to the contractor if the occurrence could reasonably have been foreseen by an experienced contractor at the time of tender, or if the event is consequent upon any omission or default by the contractor. Any such transfer of risk should reflect the intention of the employer.

2.3.6   Liquidated damages for delay to completion and pain-share/gain-share frameworks are designed in part to compensate the employer for the costs associated with the late delivery of the project and in part to incentivise the contractor to complete

---

[1] The principles of the Guide are the foundation for the CIOB's Time and Cost Management Contract (2015), which together with the Consultancy Appointment and Subcontract constitute the suite of Time and Cost Management Contracts and associated documents. It is available free from http://www.ciob.org/insight/time-and-cost-management-contract-suite (accessed 10 December 2017).

the works in the time allowed under the contract. Factors to consider in establishing a framework include:

- A reasonable contract period has to be established. This may be determined by the employer, or prospective consultants and/or contractors may be asked to pose a contract period as part of the tender process (possibly with incentives for the best time offered). An unreasonably short project duration is not a performance incentive; it will drive up costs, increase the risk of quality failures and result in late delivery.

- The right level of damages for late delivery and where appropriate incentives for early completion will incentivise the contracting party. If the liquidated damages are excessive, they may prove ineffective in encouraging timely performance, because they will be unenforceable. If the potential costs of late delivery are too severe (even if justifiable), competent contractors with an adequate workload may refuse to submit a tender or may increase their bid price to compensate for their perceived risk exposure.

- Liquidated damages imposed for late delivery or the prospect of general damages rarely compensate for the full cost of late delivery. Ultimately, neither may be recoverable for a range of reasons including, for example, the contractor going out of business.

2.3.7   An employer's time contingency should be established for the risks to be borne by the employer and, if not included in the contract period, then included as a buffer outside the contract period but within the overall project time allowed.

2.4   Project planning strategy

2.4.1   An effective planning strategy will recognise that time expires at a regular and consistent rate, from inception to completion, whether it is used effectively, or not used at all. Accordingly, effective time management will demonstrate the most effective use of available time between inception and completion, in all circumstances.

2.4.2   Major projects are necessarily conceived a long time before they are designed, designed a long time before they are constructed, and put into use a long time after construction starts.

2.4.3   Generally, the larger the project, the longer the time between inception and completion, and the more likely it is that there will be changes and other intervening events to be taken into consideration in the future.

2.4.4   A robust planning strategy which facilitates the effective management of changes to the objectives, work content, sequences, resources, and other intervening events is essential to the effective management of time on major projects.

2.4.5   The most effective time management strategy starts in the design stages of a project, so that the project is designed to be time effective without compromising out-turn cost or quality.

2.4.6   Where it is foreseeable that the occurrence of a predictable risk will delay a sequence, a time-effective planning strategy will take into account the likelihood of that risk occurring and, if the risk cannot be avoided entirely, the strategy will describe appropriate time and cost contingencies for its occurrence.

2.4.7   At inception, the employer, in conjunction with its consultants and, if appointed, the contractor, should prepare a competent appraisal of the risks that could delay the progress of the work and include contingency periods for them in the development schedule and in the contractor's working schedule. The risk appraisal should be reassessed as the design proceeds and throughout the course of the project. Appendix 1 lists the generic risks that are normally borne by the employer.

2.4.8   Where appropriate, the planning strategy should facilitate the work sequences that are likely to be severely disrupted and delayed by foreseeable events to be separated into parallel paths, rather than sequential paths. Appendix 2 contains a description of four case studies from around the world illustrating the increased time and cost risks of a single path and the reduced time and cost risks of parallel path scheduling.

2.4.9   If the potential for time management has not been considered during the design stages of a project, then the opportunities for effective management of change and other intervening events may be limited during the construction stages. The case studies in Appendix 2 illustrate some of the difficulties that can flow from failure to take into consideration the opportunity for alternative sequences of construction at the design stage.

2.4.10   In order to achieve the most effective time management strategy, the employer, design team, contractor and subcontractors must have the opportunity to contribute to the effective planning of the part, or parts, of the project with which they are concerned.

2.4.11   As part of the evolution of the planning method statement, the planning strategy is to be reviewed and articulated before scheduling commences.

2.4.12   The planning strategy is to be recorded in the planning method statement and is to be reviewed and revised regularly, particularly when intervening events not previously taken into account have occurred.

2.4.13   There must be an effective audit trail between the information in the planning method statement and the data in the relevant schedule.

2.5   Progress record strategy

2.5.1   The purpose of records is to provide the historical data that documents the timing of work carried out, the resources used and the conditions under which it was carried out. Records should be maintained for the design, procurement and construction of the works. They will include information on:

■   work done, including start and finish dates, work accomplished and, for work in progress, the time required to complete;

■   resources used and the resources required to complete work in progress;

■   quality control; and

■   information flow.

They will also include anything material to the particular type of project upon which it is essential to keep trade-specific data, for example:

■   piling operations; and

■   welding operations

and, where the work is carried out on site, the conditions under which it was executed, including the measure of:

■   wind

■   rain

■   temperature

■   humidity.

2.5.2   Effective and meaningful records are important because they are:

■   the factual evidence against which the schedule is updated with progress data;

■   proof of the expected productivity data that was used to plan the work;

■   the factual basis for the productivity assumptions for the future of the project; and

■   the basis for demonstrating the cause and effect of intervening events.

2.5.3   Effective records are those which can easily be understood, accessed, retrieved, sorted, filtered and reported upon. It is thus essential that they be kept electronically, as database records.

2.5.4   There must be an effective audit trail between the records of work actually performed by the resources actually employed and the data in the schedule.

2.5.5   Except in the most unusual circumstances, records are to be made and maintained at regular intervals. Depending upon the project, work type and types of record, this may mean monthly, weekly, daily or hourly.

2.5.6   A process is to be identified for verifying and agreeing the progress data at each update, identifying and removing any inconsistencies and archiving.

2.6   Schedule design strategy

2.6.1   The overarching purpose of a schedule is to indicate when work is to be performed in the future. The strategy for effective schedule preparation must be to provide a prediction of what is the intended timing and sequence of work yet to be carried out. In other words, it must show how the work is planned to happen and be a predictive, practical model for the future conduct of the work.

2.6.2   The preparation of a competent and effective schedule must be based upon a sound, time-effective planning strategy and must not proceed until the planning method statement has been developed.

2.6.3   The schedule objectives, structure and layout should be designed before scheduling commences. The content of the schedule will change during the development of the design and construction of the project. Accordingly, the schedule is to be designed to accommodate change in subject matter, content, method and source of data without compromising transparency between the schedule and other time-related information.

2.6.4   Because the content and timing of the work are unlikely to remain unchanged for more than a brief period, the schedule must be designed to accommodate changes. The schedule must predict the consequences of any change.

2.6.5   There must be an effective audit trail between the data on the schedule and the planning method statement, identifying why it is planned that way.

2.7   Schedule update strategy

2.7.1   The purpose of a progress update is to demonstrate the effect of progress already achieved on the timing and sequence of work still to be carried out. The sequence and timing of future work is necessarily dependent upon both the actual progress made to date and the work capable of being accomplished with the available resources in the future. Therefore, it is essential for effective time management that the schedule is updated at regular intervals.

2.7.2   The appropriate interval of schedule update should be consistent with, and should not exceed, reporting periods.

2.7.3   Each schedule update is to identify the status of the schedule by reference to a data date or 'time now' at which the progress of the work is recorded in the progress records. All work started is to show a start date earlier than (to the left of) the data date and all work to be performed in the future must be scheduled after (to the right of) the data date. No other approach is acceptable for demonstrating effects on progress of timing and sequence.

2.7.4   Where any activity has started or finished out of sequence, the logic of the schedule must be adapted to accommodate the changed sequence, and the planning method statement must be amended to demonstrate the logic which was actually followed and the reason for the change.

2.8   Schedule revision strategy

2.8.1   The content of the schedule will change during the development of the design and the construction of the project. Accordingly, the schedule is to be designed to accommodate change in subject matter, content, method and source of data without compromising transparency between the schedule and other time-related information.

2.8.2   Because the content and timing of the work are unlikely to remain unchanged for more than a brief period, the schedule must be designed to predict the consequences of any change. As better information becomes available and the project proceeds, the development of the schedule should be based on increasing the density of the schedule.[2]

2.8.3   The strategy for schedule revision must also take account of the regular review and updating of project risks. The choice of whether a risk can be avoided entirely or, if not, whether it should be accounted for by revision of the schedule (either to take that sequence out of the critical path or to allow for the risk by use of a time contingency in the sequence) will vary throughout the course of the project, as will the perception of the relative importance of the criteria driving the decision-making process.

2.8.4   Whatever choices are made and/or decisions taken, they can never be any better than the information upon which they are based at the time. It follows that if, when decisions are made, the information upon which they are based is incomplete and/or ill-defined, notwithstanding that the choices may be precisely described and the decisions recorded, they may turn out to be poor in the light of later and better information and should be open to change as necessary. For example, the strategy for schedule revision must take account of the intended sequence for development of the design as the design proceeds, and the schedule should be amended to reflect design decisions (and the design decisions should be cognisant of their effect on the schedule). In major construction and engineering projects, it is unlikely that full details of all the work to be performed will be available at tender stage, and cost contingencies, provisional sums and prime cost sums are likely to be required to be included in the tender package for work still to be designed, or decided upon during

---

[2] See 'Schedule density design' at Section 4.2.

the construction process, all of which will require time allowances and their subsequent adjustment.

2.8.5   Irrespective of whether approval of revisions is contractually required, the scheduler is to ensure that no schedule revision will constitute, or lead to:

■   an illegal operation;

■   a breach of contract;

■   a hazard to health and/or safety of personnel in the process of construction;

■   a hazard to the safety and/or stability of the permanent work, temporary work or other property;

■   a method or sequence which is not conducive to effective time control; or

■   misrepresentation, or otherwise unethical conduct.

2.8.6   Any revision to the schedule should encourage safety, contractual and legal compliance, and effective time control.

2.8.7   When a submission of a revision is required to be made for any reason, it is to be submitted as soon as the revision is complete.

2.8.8   A submission for approval of a proposed revision to the schedule is to be considered promptly and either approved (in which case the revised schedule becomes the working schedule for the future of the project) or rejected. The criteria for rejection should be limited to those set out above, and full details of the reasons for any such rejection should be given.

2.9   Time risk management strategy

2.9.1   The management of time requires a proactive assessment of those occurrences that are likely to cause an effect on the progress of the work at some point in the future. This is sometimes referred to as the risk of change, where change is defined as anything that is not planned. Change can have both positive and negative effects on the schedule.

2.9.2   Time management requires that all of the risks likely to affect the progress of the work from inception to completion are identified as early as possible, together with a meaningful strategy for dealing with the possibility of their occurrence and their likely consequences.

2.9.3   An effective risk management strategy should comprise a review of the answers to the following questions for the risk of each identified intervening event:

■   What can go wrong?

■   What is the likelihood of that happening?

■   When is it likely to happen?

■   If the risk materialises, is it likely to have a consequential effect on a key date, sectional completion date or completion?

■   Can the planning strategy be altered to avoid the possibility of the risk occurring?

■   Can the planning strategy be altered to avoid the possibility of the risk, if it occurs, having an adverse effect on progress?

■ Can the planning strategy be altered so that any delay to progress that is likely to occur is minimised or managed so it has no effect on the achievement of key dates, sectional completion dates or the completion date?

■ Is a time contingency or the adjustment of a time contingency required?

■ If not, how is the likelihood of the occurrence of the risk and/or its consequences to be managed?

2.9.4   In addition to its many other uses, the project risk register should fulfil the purpose of identifying those possible occurrences which, if they materialise, can be reasonably foreseen to affect progress, define the likely effect, and describe what can be done to minimise the occurrence of those risks having an adverse effect on key dates, sectional completion dates and on completion.

2.9.5   Once the master risk register has been set up, it must be reviewed regularly against emerging risks and, where necessary, revised by reassessment of priorities and foreseeable risks through the evolution of the design, tender and construction, including consultancy appointments and employer procurement through to testing, commissioning and occupation. Changes in the risk register should be incorporated into the schedule, and the risk reviews should consider the current schedule status.

2.9.6   There cannot be a time-effective planning strategy unless the risks which are likely to affect the future conduct of the work are allowed for, and without a planning strategy for the future conduct of the work, it is impossible to properly appraise the risks that are likely to affect it. It therefore follows that the development of an effective time risk management strategy is inseparable from a time-effective planning strategy.

2.9.7   In relation to those risks that cannot be avoided entirely, the purpose of a time risk management strategy must be to facilitate the recovery of lost time economically and effectively. This is usually achieved by identifying time contingency buffers, strategically, against the foreseeable risk of delay to progress in critical zones.

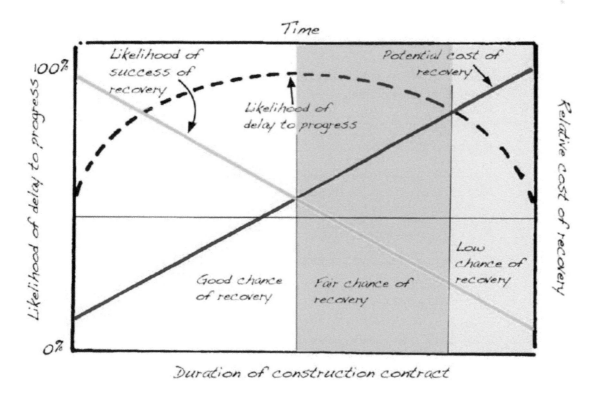

Figure 1 Profile of risk of delay to progress.

2.9.8   During the early stages of every project there should be many alternative sequences that can be pursued to optimise time performance. However, those options will inevitably reduce as decisions are taken and the project proceeds. It is unusual to find a project with many opportunities for resequencing in its later stages.

2.9.9   A typical profile of the risk of delay to the start or finish of discrete activities and the opportunity and cost escalation of dealing with their consequences can be illustrated as in Figure 1.

2.9.10   However, it is obvious that, no matter how good the risk management, it is not possible to 'manage out' the likelihood of occurrence of all risks and this should not be seen as a substitute for a time risk management strategy that can account for those risks which have a reasonable probability of maturing into intervening events. Typically, but dependent upon the contract and the project, those employer's risks for which allowances should be considered are listed in Appendix 1.

2.9.11   A shift in timing is almost certain to occur as a result of the maturation of one or more foreseeable risks into an intervening event at some time during the life of the project. On a major project it is indefensible to proceed on the basis that an intervening event will not occur or, if it does, its effect on progress (and its consequences) can be dealt with at the time; effective time management requires a proactive approach to dealing with foreseeable intervening events.

2.9.12   When an intervening event occurs, and takes up time and/or resources which do not form part of the original agreed contractual obligations, records must be maintained to identify each intervening event and the work and/or suspension of work that flows from it.

2.9.13   The work content, timing and sequence involved in responding to an intervening event are to be estimated and added to the schedule at the earliest opportunity and the schedule is to be updated with factual information as it becomes available.

2.10   Schedule quality control strategy

2.10.1   There must be an effective schedule quality control process, which is independent of any contractual third party review that may be required. The schedule quality control process is to be specifically designed to review, and maintain alignment between, the planning method statement, the schedule, each schedule revision and/or update and the record-keeping processes.

2.10.2   The purpose of the quality control process is to ensure that the recommended practices in the Guide are followed and that any departure from the recommended practices is adequately documented with:

■   the reason for such departure;

■   the acceptance by the contracting parties of the need for such departure; and

■   recognition of the likely consequences of the departure.

2.11   Building information modelling strategy

2.11.1   The rapid evolution of BIM and other computer modelling techniques is flowing through to project scheduling. The ability of the software to illustrate the intended sequence of work as a dynamic 4D model and link the schedule with many different data-sets is already available. The development of the working schedule by interactively manipulating on-screen images of the time-scaled activity network is becoming available. The combination of these technologies to permit the working schedule to be developed by directly interacting with a virtual 3D model of the project is not far away. However, regardless of the technology used to develop the schedule, the working schedule should still be developed and maintained as a DTM.

2.11.2   The benefits of 4D modelling are maximised when adopted in the early stages of a project. Starting early will integrate the use of 4D modelling within the design development process, which will assist time-related decision making, and facilitate communication both within the project team and with external third parties such as development control authorities.

2.11.3   The provision of a central resource for coordinated time and design information will help to engage all parties with an interest in the timing of the work, leading to faster, more informed decisions.

2.11.3.1   Coordinating 3D design geometry with time-related information permits both pieces of information to be viewed dynamically, in one space. The 4D model provides a single environment for design and time-related information to inform one another, and to predict and accommodate the consequences of change in the 3D virtual model. 4D modelling can assist with change risk management, by visually demonstrating in the 3D virtual model the effect of change and how different options might work.

2.11.4   4D modelling can be used:

- to visualise the 3D virtual objects comprising a scheduled activity;

- to identify in the schedule the 3D virtual objects comprising each activity;

- to visualise the scheduled sequence of activities in the 3D model;

- to help identify and assess risks and opportunities;

- to investigate the potential strategies for avoidance of a particular risk occurring and/or for minimising or avoiding the consequences of a risk that materializes;

- to save the reason for change linked or embedded to both the virtual objects and the activities; and

- to provide an audit trail of what occurred on the project as a virtual as-built model of construction progress, with linked and embedded information.

2.11.5   Like scheduling software, 4D software cannot produce a great model on its own. Software does not make decisions for the scheduler, therefore using 4D to create a useful schedule still requires fundamental skills of scheduling.

2.11.6   Whenever BIM is used in a 4D or 5D environment, care should be taken to make sure that the synchronisation and integration between the 3D model and the time data in a 4D model and/or the cost data in a 5D model does not render the modelling too data-heavy and become a distraction.

2.12   Communication strategy

2.12.1   Effective communication requires all parties to have access to a common data-set, which is easy to access, use and interpret. Effective communication is essential for the identification and management of emergent risks and for deciding between alternatives for future performance in the most timely and effective manner. This requires that all relevant persons have easy access to:

■ project scope and objectives;

■ the planning method statement explaining the strategies for the project;

■ the schedule;

■ contemporaneous progress records; and

■ the risk register.

2.12.2   The employer, design consultants, contractor, subcontractors and others responsible for carrying out work or the quality control of any work performed under the contract should have equal access to the same information regarding sequencing and timing of the work to be performed.

2.12.3   All time-related information and data are to be produced and made available electronically.

2.12.4   There must be a common denominator linking information of like type between different databases and documents (e.g. the activity ID used by the project scheduling software or 3D objects in a BIM environment).

2.12.5   A well-designed 4D model can be a highly effective communication tool for the explanation of complex sequencing issues to people who are not familiar with critical path networks. However, the attachment of a lot of project information within 4D software can make the model unmanageable, and it should not be relied upon as the only tool for coordinating all of the project information. Electronic document management systems, which can be linked to the 4D model, should be adopted for data storage.

2.12.6   Open data formats can be created from the virtual model, but the modifications such viewers permit is usually restricted and their value is thus limited. However, they do provide all parties with a common data-set of the 4D model. This can assist with communication and understanding of the schedule; however, native software licences are usually required to fully interrogate and make changes to the 4D model.

3   The dynamic time model

3.1   Introduction

3.1.1   Production and use of the dynamic time model (DTM) differs significantly from the traditional method of programming or scheduling. The following list outlines some of the differences.

- Commonly, a critical path network schedule has not been used to manage the design process and, as a consequence, the design programme or schedule has often had faulty or incomplete logic, if any at all. As a result, it has not been possible to use the design programme or schedule to manage change, predict the consequences of intervening events or manage risk during the design process. In contrast, the DTM commences with the development schedule. The development schedule must be a high-quality, fully networked schedule of the design and other preconstruction processes, that can be used to predict consequences and manage change and which subsequently may be used to inform, in whole or in part, the contractor's working schedule.

- Commonly, the programme or schedule used by most contractors to manage their work has not been required to be a high-quality critical path network, and consequently those produced have often had faulty or incomplete logic, or no logic at all. As a result, it has not been possible to use the contractually required programme or schedule to manage change, predict the consequences of intervening events, or manage risk during the construction process. In contrast, the DTM required by the Guide must be a high-quality, fully networked schedule that can be used to predict consequences and to manage change.

- The contractually required programme or schedule has not always been the product of database-driven scheduling software and has been issued on paper or PDF, without the employer and its consultants having access to the assumptions, data or calculations used to produce it. In contrast, the DTM must be the product of database-driven scheduling software with a planning method statement identifying the assumptions and calculations in it, and to which full electronic access may be given to all stakeholders and their advisers.

- Traditionally the contractually required programme or schedule has been required to describe the whole of the works at a uniform density. Typically, a high level of detail has been required by most construction and engineering contracts, either without provision for (or forbidding) revision without approval. The DTM recognises that work that has not yet been designed cannot be scheduled in anything other than outline. Accordingly, the DTM comprises activities in

Guide to Good Practice in the Management of Time in Major Projects: Dynamic Time Modelling,
Second Edition. The Chartered Institute of Building.
© 2018 John Wiley & Sons Ltd. Published 2018 by John Wiley & Sons Ltd.

different densities, depending upon the degree to which the design, resources and productivity can be ascertained. The DTM is then to be revised on a 'rolling wave' basis as more detailed information becomes available

■ Traditionally, the contractually required programme or schedule has not been used by the contractor for the purposes of managing the work on site. The contractually required programme or schedule has not always been produced before the work has started, changes to it have been discouraged, and it has been used, primarily, in disputes with the client. The contractor may have produced an overall programme or schedule for compliance with the contract, but detailed programming or scheduling of the work on site, if done at all, tended to be by use of separate short-term look-aheads, which were not embedded in the contractually required programme or schedule, were not produced in the same software, and access to them was not available to the employer and its consultants. In contrast, the DTM incorporates a transparent short-term look-ahead in the High Density part of the working schedule, to which the employer and its consultants may have access.

■ Commonly, the contractor's short-term look-ahead has not been constructed as a network, has not been resource-loaded, has not contained any productivity data, and the activity durations have been estimated on an undisclosed basis. The High Density part of the DTM, on the other hand, must be resource-loaded and the activity durations must be calculated by reference to the resources planned to be applied and the productivity they can be expected to achieve.

■ Traditionally, the contractually required programme or schedule was not required to be revised, and on major projects it tended to quickly become out of date and fall into disuse. The DTM requires that the working schedule is regularly revised on a 'rolling wave' basis, to keep it up to date.

■ Commonly, progress achieved has either been estimated subjectively, gauged against the performance to intermittent milestones, or calculated by reference to the value of the work estimated to be fit for payment. The DTM required by the Guide (which has a fully resourced network for the work to be carried out in the short term) requires as its progress records details of the resources actually used, the quantity of work actually performed, and the time period of the record.

■ If progress has been marked up on the programme or schedule at all, it has often been marked up on a 'jagged line' principle (which does not identify the effect of progress achieved on future work). On the other hand, the DTM requires that the critical path of the updated schedule is recalculated to identify the effect of progress achieved on the key dates, sectional completion dates and/ or the completion date, and to identify any corrective sequence that must be followed in order to achieve the due dates.

■ Although it is common for prime cost, provisional cost and cost contingency sums to be included in project budgets, historically no time allowance for spending such sums has been required to be included in the contractually required programme or schedule. The DTM, on the other hand, requires that appropriate time contingencies be included for all undefined work that is expected to be needed, including that arising out of perceived risks.

■ Because the actual progress of the work frequently differs significantly from the planned sequence in the contractually required programme or schedule, and it is not revised to identify additional or changed work, progress, if recorded at all, is rarely recorded in a way that is related to the contractually required programme or schedule. As a consequence, claims for extension of time and compensation have often been framed subjectively and left until after the opportunity for risk management has passed. In contrast, the DTM requires progress to be calculated, and revisions to be made regularly to keep the schedule up to date and for the effects of change to be calculated objectively by incorporating the intervening event and its effects into the DTM. This creates contemporaneous information upon which informed and calculated decisions can be made, with all stakeholders and/or their advisers having transparent access to the decision-making process.

■ Traditionally, the contractually required programme or schedule bears such little resemblance to the way the work has actually been carried out that it has no use at all after completion, save to be resurrected by one side or the other in dispute proceedings. The DTM, on the other hand, when completed, contains all the resource and productivity data of the work actually constructed and can be used to good advantage in benchmarking, to improve scheduling and cost estimating on other projects.

3.1.2   In order to set into context the recommendations of the Guide for the construction and management of the dynamic time model, the following scenario describes the creation of the development schedule through to the use of the working schedule in construction, to archiving and benchmarking.[1]

3.2   The initial development schedule

3.2.1   At the conceptual design stage (the stage at which it is known whether the project is, for example, a bridge or a school and where it will be sited), it is possible to carry out a preliminary risk assessment and to plan and schedule the design process.

3.2.2   The time each design activity is likely to take is usually calculated on the basis of the fees expected to be earned divided by the fee-earner's charge-out rate to give the number of man-hours per fee-earner. This produces a High Density schedule for the short-term lead design activities by reference to the resources to be used and the activities they are to perform, together with their value.

3.2.3   Medium Density activities can be added for other lead design activities, and for support design disciplines ('structural engineering should take about three months') based on experience. Low Density can then be added for the construction period ('it will take about 18 months to build'). The estimated value of the Medium and Low Density activities should be included in the schedule and, following a risk assessment, contingency periods added to cover foreseeable risks.[2]

3.2.4   Apart from the planned sequence, the schedule will show when any identifiable key dates, sectional completion dates, and the completion date are planned to be achieved, and if any of those are specifically required by the design brief, the dates by which they are contractually required to be achieved. It will also show any time risk contingency periods that are perceived to be needed to achieve those contractual dates.

---

[1] See Appendix 4 for a tabulated summary of the process of dynamic time modelling.
[2] See 'Risk and contingencies' at Section 4.9.

3.2.5   When completed, this schedule is saved as the development schedule DS001.[3] In summary, the development schedule, at inception, calculates the following information to be communicated and reported upon:

■   the planned sequence of work from inception to completion;

■   contingency periods for foreseeable risks;

■   all planned key dates, sectional completion dates and the completion date;

■   the critical paths to each of the key dates, sectional completion dates and the completion date;

■   the interface between differing design disciplines and construction;

■   any design work that is expected to take place during construction;

■   the sequence of work which is to be followed over the next three months or so in detail (the High Density part of the schedule);

■   the resources which are allocated to be used over the next three months or so, their planned productivity and the activities to which each resource is assigned (the High Density part of the schedule);

■   the value of each activity planned to be carried out over the next three months or so by reference to the resources allocated to the work (the High Density part of the schedule); and

■   the predicted cost of the overall project (including cost and time contingencies) and the planned spend per unit period (e.g. monthly valuation).

3.3   The updated development schedule

3.3.1   To record progress achieved and consequential changes to the schedule, DS001 is copied to DS002. DS001 is archived and the active development schedule becomes DS002. Progress records[4] of time spent on particular activities by particular resources are maintained in a database, contemporaneously.

3.3.2   At the planned update, DS002 is updated from the database records with progress achieved, identifying in relation to what was planned to be achieved (in the short-term High Density part of the schedule only) what has actually been achieved by the resources actually used.

3.3.3   After the progress data has been input, the schedule is recalculated to identify any consequential changes to sequences and any of the planned to be achieved key dates, sectional completion dates and the completion date.

3.3.4   The schedule update records the state of the progress achieved on the High Density part of the schedule at the new data date and hence the value of the work done up to the data date. The consequential changes to sequence and the planned achievement of the key dates, sectional completion dates and the completion date is calculated by the software by rescheduling the work that is incomplete or not started, taking into account the progress data.

---

[3] The numbering sequence for the development schedule (DS) starting at 001 is representational – the Guide recommends archiving a version of the schedule after each of the processes described in this section and incrementing the version so that there is a clear record of the 'before' and 'after' status of the schedule and variances can be clearly identified and attributed.

[4] See 'Progress records' at Section 5.5.

3.3.5   If the rescheduled updated DS002 reveals any adverse change in the planned sequence or change to the planned key dates, sectional completion dates and the completion date, such that the contractual dates are unlikely to be met, there must be some recovery planned to get the schedule back on track.[5]

3.3.6   Where the data of productivity actually achieved shows that the resources currently being used are insufficient to achieve the required rate of progress, the future sequences must be rearranged to be consistent with the productivity that can be achieved, and where necessary actions must be planned to increase the quantity or capability of the resources available for the work.

3.3.7   Recovery of predicted delay to the key dates, sectional completion dates and the completion date may be achieved by resequencing some later parts of the design process, absorbing the design team's time contingencies, or changes in the planned resources of work yet to be performed.[6]

3.3.8   When the necessary changes in order to plan to achieve the key dates, sectional completion dates and the completion date have been put in place, this is saved as the updated development schedule DS002. In summary, DS002 calculates the following to be communicated to the project team and reported upon:

- the amount of work done in the update period
- the resources used in the update period
- the value of work done in the update period

and, by comparison between DS001 and DS002,[7] the changes made and the consequential residual effects of any delay to progress on:

- the planned sequence of remaining work from inception to completion;
- remaining contingency periods for foreseeable risks;
- all planned key dates, sectional completion dates and the completion date;
- the critical paths to each of the key dates, sectional completion dates and the completion date;
- the interface between differing design disciplines and between those interfaces and construction;
- the process which is actually to be followed over the next three months or so (the High Density part of the schedule);
- the resources that are planned to be used and their planned productivity over the next three months or so, and how any increases in quantity or capability will be implemented (the High Density part of the schedule);
- the value of each activity planned to be carried out over the next three months or so by reference to the resources allocated to the work (the High Density part of the schedule); and
- the predicted cost of the overall project (including cost and time contingencies), all sections and the planned spend per unit period (e.g. monthly valuation).

---

[5] See 'Acceleration and recovery' at Section 5.10.
[6] See 'Acceleration and recovery' at Section 5.10.
[7] See 'Schedule comparison' and 'Variable baseline target' at Section 5.9.

3.4 Calculating the predicted effect of intervening events on the development schedule

3.4.1 DS002 is copied to DS003. DS002 is archived and the active development schedule becomes DS003.

3.4.2 If any intervening event has occurred, the effect of the event on the updated schedule must now be calculated. The intervening event calculations may be carried out individually for each event, or all intervening events in a defined period may be calculated together (the 'windows' method) depending upon what the contract requires for the proof of causation.[8]

3.4.3 It is important to consider the effect of progress achieved and any conse-quential recovery before calculating the impact of intervening events, because the effect of a known cause (the intervening event) cannot be calculated if its effect could also be the result of an unknown cause (unresolved delay to progress).

3.4.4 A comparison between the productivity planned to be achieved and that actu-ally achieved without change in the planned resources will indicate whether there has been any disruption experienced. Similarly, where productivity has shown to have ceased entirely on a particular activity will indicate whether progress has been suspended.

3.4.5 The effect of the intervening event is impacted by either suspending the rel-evant working calendar (in the case of a suspension or the effect of disruption) or (in the case of a change) adding a fragnet of the event to DS003 and rescheduling to calculate its effect on the key dates, sectional completion dates and the completion date. This is saved as the impacted development schedule DS003 (there may be several versions of this dealing with different events: DS003a, DS003b, etc., depend-ing upon what is required by the contract). In summary, DS003 calculates the follow-ing to be reported upon:

- the intervening event that has occurred, is occurring, or is likely to occur

- the effect of the intervening event (independent of the effect progress achieved)

and, by comparison between DS002 and DS003, the consequential effects caused only by the intervening event on:

- the planned sequence of work since the last update of the development schedule up to completion;

- all planned key dates, sectional completion dates and the completion date;

- the critical paths to each of the key dates, sectional completion dates and the completion date;

- the interface between differing design disciplines and between those interfaces and construction;

- the process which is actually to be followed over the next three months or so (the High Density part of the schedule);

- the resources which are planned to be used and their planned productivity over the next three months or so (the High Density part of the schedule);

- the value of each activity planned to be carried out over the next three months or so, by reference to the resources allocated to the work (the High Density part of the schedule);

---

[8] See 'Change control' at Section 5.8.

■   the predicted cost of the overall project (including cost and time contingencies), all sections and the planned spend per unit period (e.g. monthly valuation);

and hence, depending on the terms of the contract as to entitlement:

■   the extension of time, if any, to which any of the design team may be entitled; and

■   the additional cost of any of the design disciplines as a result of the intervening event.

3.5 Planning to overcome the predicted effects of an intervening event

3.5.1   DS003 is copied to DS004. DS003 is archived and the active development schedule becomes DS004.

3.5.2   If the impact of an intervening event is to predict a delay to any of the key dates, sectional completion dates and/or the completion date, some acceleration may be required by the employer, in order to recover the delay to progress.[9] Depending upon the terms of the contract, this may be by instruction or by agreement and may comprise:

■   omission of some work;

■   resequencing some later parts of the design process;

■   absorption of the employer's time contingencies; or

■   changes in the planned resources of work yet to be performed.

3.5.3   When the necessary changes required to plan to achieve the key dates, sectional completion dates and the completion date have been put in place, this is saved as the accelerated development schedule DS004. In summary, DS004 calculates the following to be reported upon:

■   the changes necessary to the planned sequence of work to absorb, in whole or in part, the predicted consequences of the intervening event

and, by comparison between DS003 and DS004, the residual consequential effects, if any, caused by the intervening event on

■   the planned sequence of work since the last update of the development schedule up to completion, after acceleration procedures have been put in place;

■   any remaining contingency periods for foreseeable risks;

■   all planned key dates, sectional completion dates and the completion date;

■   the critical paths to each of the key dates, sectional completion dates and the completion date;

■   the interface between differing design disciplines and between those interfaces and construction;

■   the process which is actually to be followed over the next three months or so (the High Density part of the schedule);

■   the resources which are planned to be used and their planned productivity over the next three months or so (the High Density part of the schedule);

---

[9] See 'Acceleration and recovery' at Section 5.10.

■ the value of each activity planned to be carried out over the next three months or so by reference to the resources allocated to the work (the High Density part of the schedule); and

■ the predicted cost of the overall project (including cost and time contingencies), all sections and the planned spend per unit period (e.g. monthly valuation)

and hence, depending on the terms of the contract as to entitlement:

■ the residual extension of time, if any, to which any of the design team may be entitled; and

■ the additional cost of any of the design disciplines as a result of the planned acceleration.

3.6 Revision of the development schedule

3.6.1   DS004 is copied to DS005. DS004 is archived and the active development schedule becomes DS005.

3.6.2   As the work performed in the first update period eats into the High Density part of the schedule, any of the Medium Density activities that are now planned to occur in the period of three months or so after the data date must be reviewed and expanded to a series of High Density activities showing the resources to be used and the task sequence that is actually intended to be followed by the various resources. A similar process should be followed, when appropriate, to expand Low Density activities occurring in the next year or so into Medium Density activities. This is called the 'rolling wave' scheduling revision process.

3.6.3   In addition to time-based changes in density, as the design matures, a better idea of what processes must be followed in an increasingly more detailed design evolution, and a better idea of what will be constructed and how it may be constructed, will emerge and should inform the development of a more detailed and realistic Medium and Low Density part of the development schedule.

3.6.4   If this revision reveals any adverse change in the planned key dates, sectional completion dates and/or the completion date such that the contractual dates are unlikely to be met, there must be some recovery planned to get the schedule back on track. This may be achieved by resequencing some later parts of the design process, absorbing time contingencies, or changes in the planned resources of work yet to be performed.[10]

3.6.5   When the necessary changes to plan to achieve the key dates, sectional completion dates and the completion date have been put in place, this revision is saved as the revised development schedule DS005. In summary, by comparison between DS004 and DS005, this calculates the following planned changes to be reported upon:

■ the revised planned sequence of work since the last impact and/or acceleration process planned on the development schedule up to completion;

■ any remaining contingency periods for foreseeable risks;

■ all planned key dates, sectional completion dates and the completion date;

■ the critical paths to each of the key dates, sectional completion dates and the completion date;

---

[10] See 'Acceleration and recovery' at Section 5.10.

- the interface between differing design disciplines and between those interfaces and construction;

- the process which is actually to be followed over the next three months or so (the High Density part of the schedule);

- the resources which are planned to be used and their expected productivity over the next three months or so (the High Density part of the schedule);

- the value of each activity planned to be carried out over the next three months or so by reference to the resources allocated to the work (the High Density part of the schedule); and

- the predicted cost of the overall project (including cost and time contingencies), all sections and the planned spend per unit period (e.g. monthly valuation).

3.7 Time management of pre-construction activities

3.7.1 Steps 1–5 are repeated sequentially at regular intervals (usually monthly) until the design process has been completed to the construction tender/bid stage. At each cycle, the update, revision or impact produces the baseline against which the effect of the next process in the cycle can be calculated. For example, development schedule DS001 has no relevance once the update schedule DS002 has been produced, other than to demonstrate, by comparison, the effect of DS002 on the first idea of how the work will progress, DS001.

3.7.2 At the appropriate stage of the design development, the planned portion of the development schedule (which will include the design team's idea of how the work will be built and any interface with continuing design during the construction process) can be used as a tender/bid document.[11] Typically, the bidding contractor will be invited to adopt or recommend changes to the construction component of the development schedule, and change or reconceive entirely the building process outlined in the development schedule to arrive at the contractor's working schedule for the construction of the works.

3.8 The initial working schedule

3.8.1 At commencement of the construction contract, it is only possible to plan and schedule the construction process in detail for the first two to three months or so. This should be done on the basis of input from the identified subcontractors, indicating the resources they intend to use and the number of man-hours, plant-hours and materials to be used per activity, together with the planned value of those activities. This produces a High Density schedule for the short term by reference to the resources to be used, the activities they are to perform and their value by unit period (e.g. monthly).

3.8.2 Medium Density activities can also be planned for work to be executed between four and nine months after commencement, sometimes by reference to identified subcontractors, or by reference to resources and productivity (where the particular work has been designed completely. Low Density activities will also be added for construction works that are expected to take place more than nine months into the future (the work for these may have been designed in full, but it is more likely

---

[11] Irrespective of whether the contractor has been engaged in the project (e.g. as the lead in a 'design and construct' arrangement), the transition from the development schedule to the working schedule and the responsibilities of the design team and the construction team should still follow the processes recommended in this Guide.

that some design work will still remain to be completed). This may include work that is the subject of a prime cost or provisional sum, or the design may be subject to change as a result of better information becoming available during the construction process and the expenditure of contingency sums. Contingency periods should also be added for contractor's foreseeable risks.[12]

3.8.3   In addition to the planned sequence of work, the schedule will show when any identifiable key dates, sectional completion dates, and the completion date are planned to be achieved and, if any of those are specifically required by the contract, the dates by which they are contractually required to be achieved. Any time risk contingency periods that are perceived to be needed to achieve those contractual dates will also be shown.

3.8.4   When completed, this is saved as the working schedule WS001.[13] In summary, WS001 calculates the following information to be communicated and reported upon:

- the planned sequence of work from commencement to completion;

- contingency periods for foreseeable risks;

- all planned key dates, sectional completion dates and the completion date;

- the critical paths to each of the key dates, sectional completion dates and the completion date;

- the interface between employer's obligations and construction;

- the interface between the design team's obligations and construction;

- the interface between third party's obligations and construction;

- the interface between subcontractors;

- the sequence of work which is to be followed over the next three months or so in detail (the High Density part of the schedule);

- the resources which are allocated to be used over the next three months or so, the productivity they are expected to achieve and the activities to which each resource is assigned (the High Density part of the schedule);

- the value of each activity planned to be carried out over the next three months or so, by reference to the resources allocated to the work (the High Density part of the schedule); and

- the predicted cost of the overall project (including cost and time contingencies) and the planned spend per unit period (e.g. monthly valuation).

3.9   The updated working schedule

3.9.1   To record progress or accommodate consequential changes to the schedule, WS001 is copied to WS002. WS001 is archived and the active working schedule becomes WS002. Progress records of time spent on particular activities by particular resources are maintained in a database contemporaneously.[14]

---

[12] See 'Risk and contingencies' at Section 4.9.

[13] The numbering sequence for the working schedule (WS) starting at 001 is representational – the Guide recommends archiving a version of the schedule after each of the processes descried in this section and incrementing the version so that there is a clear record of the 'before' and 'after' status of the schedule and variances can be clearly identified and attributed.

[14] See 'Progress records' at Section 5.5.

3.9.2   At the planned update, WS002 is updated from the database records of progress achieved, identifying the work properly executed in relation to what work was planned to be carried out and the resources actually used (in the short-term High Density part of the schedule only). The input to the update changes the data date of the schedule, and records at the new data date the state of the progress achieved in the High Density part of the working schedule.

3.9.3   After the progress data have been input, the schedule is recalculated to identify any consequential changes caused to sequences and to any of the planned to be achieved key dates, sectional completion dates and/or the completion date. The changes to sequence and the planned achievement of dates is calculated by the software by rescheduling the work, based on the added progress data against the new data date.

3.9.4   The schedule update records the state of the progress achieved on the High Density part of the schedule at the new data date, and hence the value of the work done up to the data date. The consequential changes to sequence and the planned achievement of the key dates, sectional completion dates and/or the completion date is calculated by the software by rescheduling the work that is incomplete or not started, taking into account the progress data.

3.9.5   If the rescheduled updated working schedule reveals any adverse change in the planned sequence or to the planned key dates, sectional completion dates and/or the completion date, such that the contractual dates are unlikely to be met, there must be some recovery planned to get the schedule back on track.[15]

3.9.6   Where it is plain, from the data of productivity actually achieved, that the resources currently being used are insufficient to achieve the planned required rate of progress, either the future resources must be increased or the future sequences must be rearranged to be consistent with the productivity that can be shown from the progress records to be achievable and, where necessary, actions must be planned to increase the quantity or capability of the resources available for the work. Recovery of any predicted delay to the key dates, sectional completion dates and/or the completion date may be achieved by resequencing some later parts of the construction process, changing the planned resources on work yet to be performed, or absorbing the contractor's time contingencies, in whole or in part.

3.9.7   When the necessary changes to plan to achieve the key dates, sectional completion dates and the completion date have been put in place, this is saved as the updated working schedule WS002. In summary, WS002 calculates the following to be communicated to the project team and reported upon:

- the amount of work done in the update period

- the resources used in the update period

- the value of work done in the update period

and, by comparison between WS001 and WS002,[16] the changes made and consequential residual effects, if any, of any delay to progress on:

- the planned sequence of work from commencement to completion;

- any remaining contingency periods for foreseeable risks;

- all planned key dates, sectional completion dates and the completion date;

---

[15] See 'Acceleration and recovery' at Section 5.10.
[16] See 'Schedule comparison' and 'Variable baseline target' at Section 5.9.

■ the critical paths to each of the key dates, sectional completion dates and the completion date;

■ the interface between employer's obligations and planned construction;

■ the interface between the design team's obligations and planned construction;

■ the interface between third party's obligations and planned construction;

■ the interface between subcontractors;

■ the process which is actually to be followed over the next three months or so (the High Density part of the schedule);

■ the resources which are planned to be used over the next three months or so, their planned productivity and how any increases in quantity or capability of resources will be implemented (the High Density part of the schedule);

■ the value of each activity planned to be carried out over the next three months or so, by reference to the resources allocated to the work (the High Density part of the schedule); and

■ the predicted cost of the overall project (including cost and time contingencies), all sections and the planned spend per unit period (e.g. monthly valuation).

3.10 Calculating the predicted effect of intervening events on the working schedule

3.10.1 WS002 is copied to WS003. WS002 is archived and the active working schedule becomes WS003.

3.10.2 If any intervening event has occurred since WS001 was published, the effect of it on the updated working schedule must now be calculated. The intervening event calculations may be carried out individually for each event, or all intervening events in a defined period may be calculated together (the 'windows' method) depending upon what the contract requires for the proof of causation.[17]

3.10.3 It is important to consider the effect of progress achieved and any consequential recovery before calculating the impact of intervening events, because the effect of a known cause (the intervening event) cannot be calculated if its effect could also be the result of an unknown cause (unresolved delay to progress).

3.10.4 A comparison between the productivity planned to be achieved and that actually achieved without change in the planned resources will indicate whether there has been any disruption experienced. Similarly, where productivity has shown to have ceased entirely on a particular activity will indicate whether progress has been suspended.

3.10.5 The effect of the intervening event is impacted by either suspending the relevant working calendar (in the case of a suspension or the effect of disruption) or (in the case of a change) adding a fragnet of the event to the working schedule WS002 and rescheduling to calculate its effect on the key dates, sectional completion dates and the completion date. This is the impacted working schedule WS003 (there may be several versions of this dealing with different events WS003a, WS003b, etc., depending upon what is required by the contract). In summary, WS003 calculates the following to be reported upon:

■ the intervening event that has occurred, is occurring, or is likely to occur

■ the effect of the intervening event (independent of the effect progress achieved)

---

[17] See 'Change control' at Section 5.8.

and, by comparison between WS002 and WS003, the consequential effects caused only by the intervening event on:

- the planned sequence of work from commencement to completion;

- all planned key dates, sectional completion dates and the completion date;

- the critical paths to each of the key dates, sectional completion dates and the completion date;

- the interface between employer's obligations and planned construction;

- the interface between the design team's obligations and planned construction;

- the interface between third party's obligations and planned construction;

- the interface between subcontractors;

- the process which is actually to be followed over the next three months or so (the High Density part of the schedule);

- the resources which are planned to be used over the next three months or so and their planned productivity (the High Density part of the schedule);

- the value of each activity planned to be carried out over the next three months or so, by reference to the resources allocated to the work (the High Density part of the schedule); and

- the predicted cost of the overall project (including cost and time contingencies), all sections and the planned spend per unit period (e.g. monthly valuation)

and hence, depending on the terms of the contract as to entitlement:

- the extension of time, if any, to which the contractor and any of its subcontractors may be entitled; and

- the additional cost, if any, as a result of the effect of the intervening event and its effect on the time-related costs (preliminaries/site costs).

3.11 Planning to overcome the predicted effects of an intervening event

3.11.1 WS003 is copied to WS004. WS003 is archived and the active working schedule becomes WS004.

3.11.2 If the impact indicates a predicted delay to any of the key dates, sectional completion dates and/or the completion date, some acceleration may be required by the employer in order to recover the predicted delay to progress caused by the event at the employer's risk.[18] Depending upon the terms of the contract, this may be by instruction or by agreement and may comprise:

- omission of some work;

- resequencing some later parts of the work;

- absorption of the employer's time contingencies; or

- changes in the planned resources of work yet to be performed.

3.11.3 When the necessary changes to plan to achieve the key dates, sectional completion dates and/or the completion date have been put in place, this is saved

---

[18] See 'Acceleration and recovery' at Section 5.10.

as the accelerated working schedule WS004. In summary, WS004 calculates the following to be reported upon:

■ the changes necessary to the planned sequence of work to absorb, in whole or in part, the predicted consequences of the intervening event

and, by comparison between WS003 and WS004, the residual consequential effects, if any, caused by the intervening event on:

■ the planned sequence of work from commencement to completion;

■ any remaining contingency periods for foreseeable risks;

■ all planned key dates, sectional completion dates and the completion date;

■ the critical paths to each of the key dates, sectional completion dates and the completion date;

■ the interface between employer's obligations and planned construction;

■ the interface between the design team's obligations and planned construction;

■ the interface between third party's obligations and planned construction;

■ the interface between subcontractors;

■ the process which is actually to be followed over the next three months or so (the High Density part of the schedule);

■ the resources which are planned to be used over the next three months or so and their planned productivity (the High Density part of the schedule);

■ the value of each activity planned to be carried out over the next three months or so, by reference to the resources allocated to the work (the High Density part of the schedule); and

■ the predicted cost of the overall project (including cost and time contingencies), all sections and the planned spend per unit period (e.g. monthly valuation)

and hence, depending on the terms of the contract as to entitlement:

■ the residual extension of time, if any, to which the contractor and any affected subcontractors may be entitled;

■ the predicted prolongation costs, if any, to which the contractor and any affected subcontractors may become entitled; and

■ the additional cost to the employer as a result of the acceleration.

3.12 Revision of the working schedule

3.12.1 WS004 is copied to WS005. WS004 is archived and the active working schedule becomes WS005.

3.12.2 As the work performed in the first update period eats into the High Density part of the schedule, any of the Medium Density activities that are now planned to occur in the period of three months or so after the data date must be reviewed and expanded to a series of High Density activities showing the resources to be used and the task sequence that is actually intended to be followed by the resources. A similar process should be followed when appropriate to expand Low Density activities occurring in the next year or so into Medium Density activities. This is called the 'rolling wave' scheduling revision process.

3.12.3   In addition to time-based changes in density, as the work progresses a better idea of what processes must be followed in an increasingly more detailed design and a better idea of what will be constructed and how it may be constructed will emerge, and should inform a more detailed and realistic Medium and Low Density part of the working schedule.

3.12.4   If this revision reveals any adverse change in the planned key dates, sectional completion dates and the completion date such that the contractual dates are unlikely to be met, there must be some recovery planned to get the schedule back on track.[19] This may be achieved by resequencing some later parts of the construction process, absorbing time contingencies, or changes in the planned resources of work yet to be performed.

3.12.5   When the necessary changes have been put in place to show how completion of the relevant work by the key dates, sectional completion dates and the completion date is planned to be achieved, this revision is saved as the revised working schedule WS005. In summary, by comparison between WS004 and WS005, this calculates the following planned changes to be reported upon:

- the revised planned sequence of work since the last impact and/or acceleration process planned on the working schedule up to completion;

- all planned key dates, sectional completion dates and the completion date;

- any remaining contingency periods for foreseeable risks;

- the interface between employer's obligations and construction;

- the interface between the design team's obligations and construction;

- the interface between third party's obligations and construction;

- the interface between subcontractors;

- the process which is actually to be followed over the next three months or so (the High Density part of the schedule);

- the resources which are planned to be used over the next three months or so and their planned productivity (the High Density part of the schedule);

- the value of each activity planned to be carried out over the next three months or so, by reference to the resources allocated to the work (the High Density part of the schedule); and

- the predicted cost of the overall project (including cost and time contingencies), all sections and the planned spend per unit period (e.g. monthly valuation).

3.13   Continuing time management of construction activities

3.13.1   Steps 1–5 are repeated sequentially at regular intervals (usually monthly) until the completion of the construction contract.

3.13.2   At each cycle, the update, revision or impact analysis produces the updated baseline against which the effect of the next period of progress in the cycle can be calculated. For example, working schedule WS001 has no relevance once the updated working schedule WS002 has been produced, other than to demonstrate, by comparison, the effect of the update WS002 on the first idea of how the work will progress.

---

[19] See 'Acceleration and recovery' at Section 5.10.

3.13.3   At the appropriate stage of construction, the remaining portion of the working schedule (which will include any interface with continuing design during the construction process and the intended fit-out schedule) can be developed as an occupational commissioning schedule.

3.13.4   Typically, the commissioning engineers and the employer will have their own ideas about the sequences they wish to follow to test, fit out and to bring the engineering aspects into use and to train their employees in use of the project, but unless this is to be left until after the completion of the construction contract (which would be very unusual), it is normally planned for at least six months before planned completion of construction, and will then be planned against the projected planned completion dates of the work contained in the latest revision of the working schedule.

3.13.5   The last update should complete the as-built schedule as a high density record of the sequence in which the works were actually constructed, the resources actually used and the productivity actually achieved.

3.14   Benchmarking

3.14.1   At every update of the schedule, the data captured can be analysed and refined to produce a databank of achievable productivity, in identified activities, in a variety of experienced working conditions. This data can be used to refine and render more accurate assumptions used in the development and/or working schedule and also to provide better information for planning, scheduling and cost-estimating other projects.

4 Developing the dynamic time model

4.1 Introduction

4.1.1   In any project, there will be many parties with a legitimate interest in ensuring that the timing of the work is managed effectively. Typically, these will include:

- those financing the project
- the employer
- the contractors
- the subcontractors
- the suppliers
- the design consultants
- the project manager
- the contract administrator
- the time manager
- the cost manager

4.1.2   The purpose of the time model is to indicate when in the future and in what sequence the planned work is to be performed, and how key dates, sectional completion dates and the completion date are to be achieved. The time model can also be updated to show the consequences of any changes or departures from the plan; the variance can be measured and the effect predicted, communicated and managed.

4.1.3   At any point in time, the time model can only be as accurate a prediction of the future as the state of current knowledge will permit. Accordingly, it must be conceived as a model that can be improved upon as better information becomes available or circumstances change.

4.1.4   In order to facilitate efficient time management, the time model should be constructed so as to differentiate between work that can be predicted:

- in outline in the long term (Low Density)[1]
- in detail but with some information missing in the medium term (Medium Density), and
- accurately in the content, sequence and resources to be employed on work which will be carried out in the short term (High Density).

[1] See 'Schedule density design' at Section 4.2.

Guide to Good Practice in the Management of Time in Major Projects: Dynamic Time Modelling,
Second Edition. The Chartered Institute of Building.
© 2018 John Wiley & Sons Ltd. Published 2018 by John Wiley & Sons Ltd.

4.2   Schedule density design

4.2.1   Unless the work is designed in its entirety and all subcontractors and specialists are appointed before any work commences, at the outset it is unlikely to be possible to plan the work in its entirety at a high level of detail.

4.2.2   However, if time is to be managed effectively, before work commences the activities to be carried out in the near future must be identified in detail, together with the resources to be applied and their expected productivity.

4.2.3   To accommodate these requirements, the density of the schedule should increase from that which is possible and necessary for feasibility purposes and long-term planning to the level of detail needed for current work activities, as better and more certain information becomes available. A typical profile of schedule density against predictability is illustrated in Figure 2.

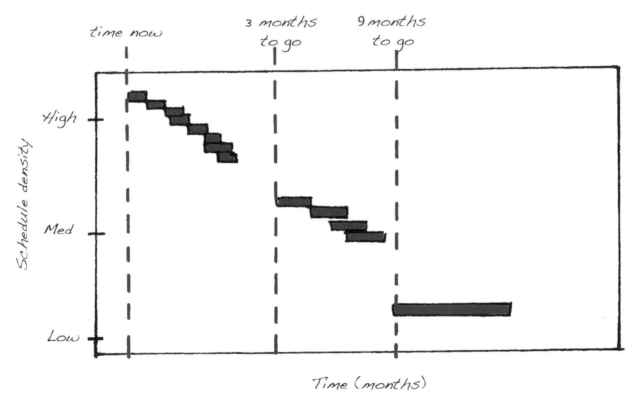

Figure 2 Graph of schedule density in relation to predictability.

4.2.4   The requirements of different densities of scheduling for different purposes must be taken into consideration at the schedule design stage and should be defined in the planning method statement.[2]

4.2.5   A profile of an activity, as it varies from Low Density through Medium to High Density, is illustrated in Figure 3.

4.2.6   The relationship between the Low Density, Medium Density and High Density parts of the schedule can be conveniently illustrated by taking a Low Density activity and developing it through Medium and High densities. The diagram in Figure 3 illustrates Activity B in Low Density taking 25 time units, which in Medium Density is represented by activities B1 to B4, and in High Density by B11 to B48.

---

[2] See 'Planning method statement' at Section 4.29

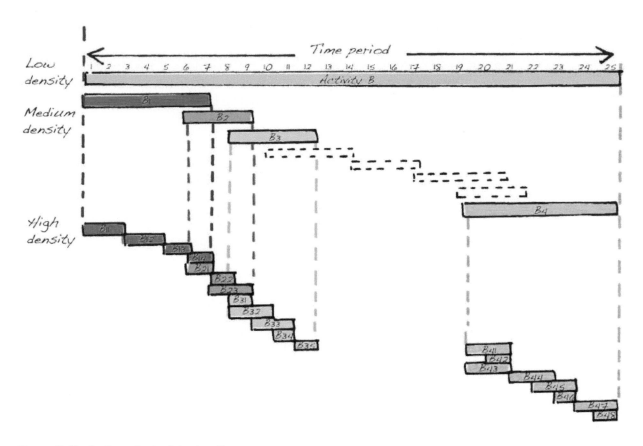

Figure 3 Illustration of schedule density.

Scheduling at Low Density

4.2.6.1   Low Density is appropriate for work which is intended to occur nine or more months in the future. Depending upon the purpose for which the schedule is intended, activities may reasonably be no more than the proposed duration of one building type, amongst others, or be trade grouped into such descriptions as 'mechanical and electrical services' and may conveniently be several months in duration.

4.2.6.2   When initially developing a schedule, it is good practice to construct a Low Density schedule for the entire project. Once this overall plan is agreed, the medium-term works can then be expanded to Medium Density, in consultation with contractors and suppliers. Then the near-term work can be expanded to High Density, in consultation with the resource teams actually responsible for the work.

4.2.6.3   Typically, standardised layouts must be prepared, usually as a bar chart or linked bar chart, to illustrate such features of the works as:

■   the periods available for finance approval, licensing and permissions, design procurement and construction;

■   the order and sequence of construction of different buildings, site works and civil engineering.

Scheduling at Medium Density

4.2.6.4   Medium Density is appropriate for work that is intended to take place between three and nine months into the future. At this stage the work should be designed in sufficient detail to be allocated to contractors or subcontractors for pricing.

4.2.6.5   Activities may reasonably be grouped into trade activities in locations with durations not exceeding two months.

4.2.6.6   Taking the same trade example, at this density the electrical services should be identified separately from the mechanical services, and the work to be undertaken by both service contractors should be identified by area and zone.

Scheduling at High Density

4.2.6.7   High Density scheduling is a prerequisite for work that is intended to take place in the short term, say within the next three months; it is at High Density that the work in progress will be recorded, monitored and reported upon.

4.2.6.8   At this stage the work should be designed in detail, the sequence and intended progress of the work clarified, and the resources and their productivity identified.

4.2.6.9   At this level the activity duration should be related to discrete activities identified by a limited area and be no greater in duration than the period against which progress is reported.

4.2.6.10   Typically, standardised layouts must be prepared to illustrate such features of the works as:

- the detailed sequence to be followed by each resource for the identified period;

- the status of the work attributed to each resource at reporting periods;

- the relationship between resources and the productivity planned and achieved;

- the relationship between costs incurred and the productivity achieved; and

- the relationship between costs incurred and costs paid.

4.2.6.11   Whether reports are to be issued as electronic data files or as hard copy will need to be established, and if in hard copy, the preferred paper size will constrain the reporting layout options available.[3]

4.3   Planning method statement

4.3.1   The planning method statement is the written description of the time management strategy, the planning logistics and the scheduling calculations and assumptions for the various parts of the works. The planning method statement is used to facilitate control of the operations and to ensure that all concerned have a clear understanding of why the work is planned and scheduled in the way it is.

4.3.2   Once the risk assessment of the project has been carried out and ways to manage the uncertainties determined, the planning method statement will set out the reasoning behind the approach to the various phases of construction and list the work encapsulated in the schedule activities forming the basis of activity logic.

4.3.3   The longest path to each key date, section, phase or project completion date should be described in a short Section, and a summarised version illustrated, either as an extract from the scheduling software, pictorially or, if produced in 4D in a BIM environment, with the help of a 3D virtual model.

---

[3] See Part 6, 'Communicating the dynamic time model'.

4.3.4   Where the working schedule is conceived as a 4D model in a BIM environment, it can be used to validate the sequences and to assist with communicating the planning method statement, so all project team members have a clear understanding of why the works are planned and scheduled in the way they are.

4.3.5   The planning method statement must be regularly reviewed, revised and updated, together with the schedule. The detail of the planning method statement should be consistent with the density of the schedule.

Planning method statement at Low Density

4.3.5.1   At this level, the planning method statement can be expected to contain:

■   a description of the work to be carried out, including design, procurement and development strategy and constraints;

■   third party and neighbour interests and interfaces;

■   description of the approach to risk assessment and the risks identified (e.g. work affecting river crossings in the fish breeding season, earthworks in winter and weather cycles to which any resource is sensitive);

■   an assessment of contingencies to be allowed for those risks;

■   a description of the activities contained in the schedule by reference to their activity ID codes;

■   the work breakdown structure;

■   calendars for working weeks and holiday periods;

■   generic resources anticipated and anticipated resource constraints;

■   permits and licences required and the decision periods expected in relation to each application and their dependencies;

■   material and equipment restrictions and availability;

■   the approach to utilities and third-party projects, licences and restrictions such as power outages;

■   the approach to schedule review, revision and updating;

■   activity content codes applied;

■   activity cost codes applied;

■   details of the phasing and zonal relationships of the project;

■   principal methods of construction;

■   details of major plant requirements;

■   site management, logistical assumptions and site welfare, temporary works including scaffolding, access and traffic management;

■   health and safety;

■   environmental considerations;

■   principal methods of procurement and their effects;

■   the methods used to estimate durations;

■   the assumed sequencing logic and an explanation of any logical constraints;

■ description of the critical and near-critical paths to key dates, sectional completion dates and completion of the works as a whole; and

■ reporting formats, communications strategy and information format.

Planning method statement at Medium Density

4.3.5.2   At this level, detail will be added to the outline information given in the Low Density schedule, including any additions, deletions, amendments or refinement of that information. The Medium Density method statement may be developed incrementally, but should always be complete before each section of the schedule is upgraded from Low Density to Medium Density. In addition to the Low Density information, it should also include:

■ identified specialist contractors, subcontractors and suppliers;

■ a key trade interface management strategy;

■ a design and procurement interface management strategy;

■ limited possessions;

■ planned overtime;

■ temporary traffic diversions and plant maintenance down-time;

■ resources anticipated and any anticipated resource constraints;

■ material and equipment restrictions and availability;

■ utilities and third-party projects, licences and restrictions such as power outages;

■ schedule review, revision and updating;

■ the methods used to estimate durations; and

■ details of plant requirements and their assumed productivity, down-time and maintenance.

Planning method statement at High Density

4.3.5.3   At this level, the planning method statement will be refined to detail the activities to be carried out in the short term, including any additions, deletions, amendments or refinement of the Medium Density planning method statement. The High Density method statement may be developed incrementally, but should always be complete before each section of the schedule is upgraded from Medium Density to High Density. In addition to the Medium Density information, it will also include a definition of:

■ resources to be employed;

■ productivity quotients expected;

■ detailed calculations of activity duration;

■ detailed methods of construction; and

■ details of plant requirements and their productivity, down-time and maintenance.

Documentation of corrections

4.3.5.4   Any corrections required to facilitate the schedule's use as a dynamic time model must be justified and documented in a corrections log. Every required correction should be documented clearly and succinctly and, when completed, the corrective action taken should also be logged for review.

4.3.5.5   The effects of any corrections, and any revisions required as a result of those corrections, should be identified in a revision to the planning method statement.

4.4   Software considerations

4.4.1   At its lowest level the software available for scheduling may be no more than a drawing tool or, at its highest, a complex arrangement of customisable databases with a graphical user interface at the front end. However, in order to be capable of producing a schedule that can perform as the time model required by the Guide, the software must have an adequately functional database at its core. This is important because the software has to be capable of computing the consequences of change; a drawing tool which simply illustrates the decisions made by the drafter cannot perform this core function.[4]

4.4.2   No matter how high the quality of the software, it cannot produce a high-quality output of its own accord. Project scheduling software, even of the highest quality, will not ensure the competent management of time. The best that software can achieve is pointing the scheduler in the direction of good practice.

4.4.3   Many software manufacturers provide extremely important and useful training focused on the features of their software and how they can be used. However, this form of training should not be confused with training in time management, nor should it be considered as a substitute for it. By analogy, many of us have experience of securing a good grounding on how MSWord™ works,[5] but even with its spell-checker and grammar-checker the software will not guarantee that what is written is useful, technically accurate, or even intelligible.

4.4.4   Whilst, from time to time, every company considering software products will wish to take into consideration matters peculiar to themselves, or matters peculiar to the project upon which they wish to work, there are certain considerations that should transcend subjective preferences, and there are certain software attributes that are desirable for the purposes of competent time management.

4.4.5   Because software changes by the day, as 'new and improved'[6] products are brought to the market, those attributes that are desirable only for the purposes of time management, irrespective of whether they are currently available in any particular product, are listed in Appendix 6.

4.4.6   It is unhelpful if different parties to a particular project use different software because different products work in different ways and, even if given the same data, may produce different results from their calculations because of the different algorithms used. Accordingly, all parties to a project should use the same software with the same settings and a departure should not be permitted.

4.4.7   Whilst getting to grips with unfamiliar software may be tedious, a competent scheduler who is provided with the recommended software training and some time to practise will be able to use the system and be sufficiently capable reasonably quickly. Unfamiliarity with scheduling software products should thus not be a serious consideration in product selection.

---

[4]   See Appendix 6, 'Desirable attributes of scheduling software'.
[5]   A product of Microsoft Corporation used for word processing.
[6]   Unfortunately this is a term often confused with more bells and whistles.

4.5 The structure of the schedule

4.5.1   There are a number of structural matters that must be given consideration and upon which decisions need to be made before scheduling commences. Some of the key constraints that may limit scheduling options may include:

■   project scope and objectives

■   resources and labour, equipment and material constraints

■   permits and licences

■   utilities and third-party projects

■   calendars

4.5.2   Other decisions which will significantly impact the design of the schedule, but which are amenable to change, if necessary, are:

■   work breakdown structure

■   activity identification coding

■   work type definition

■   density design

■   calendars

■   resources

■   licences and permissions

■   third-party projects

■   contingencies

■   activity content code

■   activity cost code

■   reporting

■   review, revision and updating the schedule

■   impacting intervening events.

4.6 Schedule types

4.6.1   Theoretically, there is no limit to the number of different types of schedule that can be produced and, historically, there has been an understandable tendency to devise a new ad hoc type of schedule for every different purpose. However, this approach should be avoided.

4.6.2   Typically, there should be no more than five schedules. The element that distinguishes between these schedules is the parties who participate in their creation. Every purpose to which they are to be put should be satisfied by organisation and filtering the schedule information relevant at the time. The schedule types are:

■   development

■   tender

■   working

- occupational commissioning

- as-built

The Development schedule

4.6.2.1   This is the schedule developed and used by the employer and its design and other consultants, usually before any contractor is involved in the project. Its focus should be the work needed to be undertaken by the employer leading into the start of the construction contract and in support of the contractor, including the design work and obtaining the approvals necessary for the contract to proceed. The work of the contractor and occupational commissioning should be included at an appropriate level of detail. The schedule should follow the same rules as to density as any other schedule. (For example, at inception the anticipated construction period for the works on site may be represented, at Low Density, by a single activity; however, by the time the scheme design has evolved, much of the anticipated construction process will be identified in Medium Density).

Tender schedule

4.6.2.2   This is the first schedule produced by the contractor prior to the award of the contract. It will incorporate any information required by the design team (transferred from the development schedule) and illustrate, at a mixture of Low and Medium Density, the sequence in which the contractor intends to achieve the contracted work and by when.

4.6.2.3   If the tender schedule is developed as a 4D model in a BIM environment, it can be used to illustrate the order in which the design team have conceived that the work may progress.

Working schedule

4.6.2.4   This is a refined and expanded version of the tender schedule and will be the schedule used for planning and progressing work on site from commencement to completion. The working schedule will map all necessary interfaces with the development schedule and the occupational commissioning schedule to provide horizontal traceability through all of the project's schedules. It must be completed in High Density for the first three months of the project, before work on site commences, and thereafter the High Density section is extended after each update to maintain the three-month High Density 'look-ahead'.

4.6.2.5   In the same way that the contractor's tender schedule is expected to be derived from the development schedule, each subcontractor's tender schedule will be derived from either the development schedule or the contractor's working schedule (depending upon the timing of the tender). After the subcontract has been agreed, the subcontractors working schedule should be developed and incorporated into the working schedule.[7]

4.6.2.6   If the working schedule is developed as a 4D model in a BIM environment, it will assist in the visualisation of workflows, optimisation of resources and the practicality of constructional sequences. It will also contribute to an understanding of challenges to be met using the 3D virtual model.

---

[7] See 'Schedule integration' at Section 4.8.

Occupational commissioning schedule

4.6.2.7   This is the employer's schedule showing how it will occupy and use the building or facility. It will contain details of partial possessions, furnishing, commissioning and testing, access and training. The schedule will usually be prepared by the employer or its consultants from the working schedule, and reviewed and revised in the light of changes made during the course of construction.

4.6.2.8   As with the working schedule, if the occupational commissioning schedule is developed as a 4D model in a BIM environment, it will assist in the visualisation of testing, commissioning and occupational workflows, optimisation of resources, and the practicality of testing and occupational sequences through the use of the 3D virtual model.

As-built schedule

4.6.2.9   This is the final schedule to be completed. It will evolve through the course of the project as the work is executed and activities are started, progressed and completed in the schedule database. The last update should complete the as-built schedule as a High Density record of the sequence in which the works were actually constructed, the resources actually used and the productivity actually achieved. This can be used for benchmarking productivity in order to improve scheduling on future projects.[8]

4.6.2.10   If the as-built schedule is developed as a 4D model in a BIM environment, it will assist in the visualisation of the construction process actually adopted using the 3D virtual model.

4.7   Schedule design

4.7.1   The schedule is the means by which the strategy and sequencing of the full scope of work to be accomplished is described. The scheduler must consider the type of schedule required, its intended content, layout, appearance, how it is to be integrated with other schedules and the nature of any reports likely to be necessary before commencing scheduling.

4.7.2   The purpose of schedule design is to set down the policy for rendering the project plans and schedules as meaningful and transparent as possible.

4.7.3   Whilst every project will have its own determining characteristics, the tabulated considerations below identify those general matters which ordinarily should be considered in designing the schedule.

4.7.4   Work that can be carried out in defined sequences, over brief periods, requires a different approach to time management from work containing activities that may take several months to complete. The former can be managed by reference to the activity start, work in progress and completion date, whereas the latter can only be managed from day to day by reference to the applied resources and productivity achieved.

4.7.5   Resource-based planning is necessary wherever productivity is more likely to affect completion than logical sequence. Typically this can be applicable to activities such as earth moving in large-scale projects, piling on open sites, pipe welding on large process plants, and most linear projects such as pipelines and tunnelling. In this form of time management, the unit productivity of each resource is interpolated

---

[8] See 'Feedback and benchmarking' at Section 6.6.

as the works proceed to provide the data for calculating the time that the activity will take to complete.[9]

4.7.6   Conversely, activity-based scheduling, with a defined sequence of working, is appropriate where the nature of the project dictates that work has to be accomplished in a logical sequence and the flow of resources through particular spaces has to be managed.

4.7.7   A key decision that must be made early in the scheduling process is determining which approach is best for each of the major sections of a project, and if different approaches are warranted in different areas, how the schedules will be integrated.

Design	**Risk**
Design responsibility	Risk allocation
Complexity of design	Employer's time contingencies
Specifications	Contractor's time contingencies
Bills of quantities	**Other contracts**
End-user requirements	Separate contractors
Procurement	Employer's contractors
Procurement strategy	Nominated subcontractors
Procurement schedule	Nominated suppliers
Materials' delivery and storage	Provisional and prime cost sums
Labour and plant resources	**Other parties' interests**
Employer-supplied goods and materials	Adjoining owners' rights
Construction	Utilities and statutory undertakings
Working conditions	Third party issues
Working hours and holidays	Licences and permissions
Logistics	**Scheduling**
Construction philosophy	Schedule content
Method of construction	Updating
Temporary works	Revision
Temporary traffic arrangements	Reporting
Sequence of construction	Completion date
Environmental	Sectional and key completion dates
Environmental conditions	Other milestones
Noise restrictions	Access, egress and possession dates
Health and safety	Information release dates
Local regulations	Submission and approvals dates

4.8   Schedule integration

4.8.1   Ideally, the development or working schedule should be created, managed, updated and revised by a schedule owner. However, major projects nearly always require a method of dealing with a part of the total project, or controlling the total project whilst the detail of specialised works is scheduled by others, and there may be multiple stakeholders, each of which is concerned with a different part of the project.

4.8.2   In most cases construction will involve a number of specialist contractors, subcontractors and suppliers. Design work may also be carried out by a number of design consultants, and some specialist contractors may also have design

---

[9] See 'Resource scheduling' at Section 4.20.

responsibilities. Where the project is delivered across a fragmented supply chain, using a number of separate contracting parties, the management of the work of the separate contracting parties creates special problems in time management.

4.8.3   The method of integration of the schedules of others and their transparency to the working schedule are important considerations in its design. The maintenance of incompatible and/or independent schedules by other contractors, subcontractors, suppliers or designers is inconsistent with effective time management.

4.8.4   The integration of the work of others in the working schedule can be achieved in a number of ways, each with its own advantages and disadvantages either to the contractor or the party whose work is to be drawn into the working schedule. In principle, schedule integration can be achieved by:

- schedule subcontracting

- master schedule and subproject

- milestone management.

Schedule subcontracting

4.8.4.1   Where preferred design consultants or subcontractors do not have the necessary scheduling ability, they should either be encouraged to employ specialists to assist them or subcontract their scheduling requirements to the lead consultant or managing contractor. In this way the lead consultant or managing contractor can prepare the other parties' scheduling contribution to their detailed requirements and incorporate it into the development schedule or the working schedule, as the case may be.

4.8.4.2   The disadvantage of this is that the lead consultant or managing contractor will need to employ sufficient schedulers to manage the totality of the works, many of them with particular specialist skills.

4.8.4.3   On the other hand, this method does keep the schedule as a single file, which is always likely to be faster to read, write and review than electronically linked multiple project files. This method also promotes a greater understanding in the lead consultant or managing contractor of the other parties' work and working methods. It also facilitates complete control over the schedule from commencement to completion. This can be of considerable value when decisions have to be made as to how sequences and interfaces may be changed to accommodate recovery or acceleration.

Master schedule and subproject

4.8.4.4   Some scheduling software will recognise a part of the schedule as a subproject, or a sub-subproject within a subproject. The idea of this is that the subproject can be viewed and/or summarised or revised and/or updated in its entirety in the sub-subproject or at a higher level in the subproject or the master development or working schedule, by being synchronised electronically.

4.8.4.5   Some software products will permit different users to have different access privileges on an activity by activity basis or, more broadly, on a work breakdown structure (WBS) or other coded basis. Granting the parties joint limited read/write access to their part of the working schedule may permit the parties to write their own scheduling contribution in whole or in part, and/or to make their own revisions and updates to the part within their access rights, whilst maintaining complete transparency and access for the lead consultant or managing contractor.

4.8.4.6   The advantages of master schedule and subproject process lie in the simplicity and privilege of access whilst maintaining control, visibility and transparency.

Milestone management

4.8.4.7   This requires the working schedule to be created to be capable of being filtered as a milestone schedule, containing only the 'give' and 'get' milestones electronically linked to the 'give' and 'get milestones in each party's schedule.

4.8.4.8   Where interfaces are managed using 'give' and 'get' milestones, these must be identified in the working schedule at appropriate points (the former represents the completion of a deliverable or other enabler, sufficient to allow following works by another party to commence). Where the 'give' milestone is linked to a 'get' milestone in another part of the schedule, the 'get' milestone represents the starting point for the second party's efforts.

4.8.4.9   The presence of these milestones allows for simple communication of key points in the project, by extracting these milestones into project reports (milestone charts).

4.8.4.10   The management of interfaces in this environment is crucial. The same principle of 'give' and 'get' milestones applies, but the links between these milestones may be held in a larger database, or even in a separate interface schedule that collects and collates the 'give' and 'get' milestones from a number of schedules, coordinating them by means of a rigidly prescribed activity numbering (or coding) scheme.

4.8.4.11   This approach can work well so long as every contracting party achieves its 'give' milestone. When it does not, and hence there is a delay to the 'get' milestone for other contracting parties, the economic management of the delaying effect (other than by absorbing contingency time) is difficult, if not impossible. For this reason, managing multiple subprojects by means of milestones alone is not recommended by the Guide.

4.9   Risk and contingencies

4.9.1   Typically, there will be many risks to be considered, some of which may require a contingency period. Much has been written in guides to project management upon the subject of the quantification of risk. It is not the intention of this publication to reproduce that, but further reference could be made to:

- Project Management Institute's *Guide to the Project Management Body of Knowledge*, 6th ed.[10] (*PMBOK® Guide*), and

- Association of Project Management's *Project Risk Analysis and Management Guide*, 2nd ed.[11] (*PRAM Guide*).

4.9.2   Only that party who is contractually liable for the consequences of the risk maturing[12] can properly determine the quantity and distribution of the contingency it perceives to be required. Accordingly, contracts should (and generally do) make clear who is contractually liable for the consequences of the risk maturing and thus who owns any time and/or cost contingency provided for that risk.

4.9.3   Reference should be made to the contract in question as to which party is to bear which risks. Typically, those matters generally at the risk of the employer, for which it may be appropriate to allocate a time contingency for the likelihood of time

---

[10] https://www.pmi.org/pmbok-guide-standards/foundational/pmbok/sixth-edition

[11] https://www.apm.org.uk/book-shop/project-risk-analysis-and-management-guide-2nd-edition/

[12] In an ideal world, this is also the party most able to manage it.

being taken up on one or more chains of activities, are listed in Appendix 1. Special risks that derive from the particular contract type and conditions of operation may also require contingency periods to be provided for.

4.9.4   Contingency planning is the calculation of the period of time needed to offset the effects of intervening events which could possibly occur, and if they occur will interfere with the smooth running of the project, at some time between inception and completion. A contingency is a planned allotment of time, which may be taken up to accommodate the occurrence of an intervening event.

4.9.5   In the same manner that cost budgets usually have an allocation of funding called 'a contingency sum', which the employer may rely upon to spend against unforeseen additional work, the schedule must have strategically placed contingency activities to absorb the time effect of the intervening events which are at the employer's risk under the contract; the employer will specify the amount of time required.

4.9.6   A prudent contractor will also make allowances for the risks it bears in the management and distribution of the resources, their variable productivity in differing circumstances, the quality of the work it carries out and any other time-related risk it carries under the contract.

4.9.7   Some scheduling software will offer a choice as to whether activities are, by default, to be scheduled as early as possible, or as late as possible. Others default to one or the other. Where an activity would otherwise be scheduled as late as possible, the introduction of a contingency period buffering its end date will have the effect of scheduling the planned commencement of the activity earlier than would otherwise be the case. The effect of this will allow for any delay in the completion of the activity to first be absorbed by the contingency period before affecting other elements in the schedule.

4.9.8   Designated non-working periods such as industry-related or statutory holidays, or weekends, are not contingency periods and should not be treated as such.

4.9.9   Contingency periods should be designed to be identified separately for the employer's risks, the contractor's risks, and for those risks that are related to specific events such as:

■   an activity or chain of activities;

■   contractor, subcontractor, supplier or other resource;

■   an access or egress date, or date of possession, or relinquishment of possession;

■   the works, any defined section, and any part of the works.

Contingencies at Low Density

4.9.9.1   At the lowest level of density, contingencies are likely to be the longest in order to provide some accommodation for the unknown aspects of the schedule. Because of the absence of precision at this level of density, the separately allocated contingencies to one party or the other may both be arrived at by applying a formula to either adjust individual activity durations or calculate the duration of a buffer activity.

4.9.9.2   One way of identifying contingencies at this level is to use a modelling approach such as Monte Carlo analysis to assess the level of contingency needed to offer an acceptable probability of completion, after allowing for both the variability in each activity duration estimate and any included risks. The Monte Carlo algorithm randomly generates values for each variable from within a predetermined range, using a predetermined distribution,[13] and calculates the schedule using these values. The process is repeated hundreds of times with different values selected from within the range to determine the probability of completing by any particular date, and as a consequence the amount of contingency that needs to be added to the critical path duration to provide the desired level of certainty. Projects are typically modelled to achieve levels of certainty ranging between a 75% (P75) and 90% (P90) probability of completing by a particular date; obviously the amount of contingency required to achieve a P90 level of certainty is much greater than a P75 level of certainty.

4.9.9.3   In order for a Monte Carlo simulation to arrive at the appropriate contingency margin, a range of dates or durations must be assigned to each scheduled activity and to each risk event in the model. From within those ranges, for each iteration, the mathematical model randomly selects a value for the duration of each activity based on the distribution. The range is usually defined as the 'optimistic (or minimum) duration', the 'most likely duration' and the 'pessimistic (or maximum) duration', where the optimistic duration is the shortest that can reasonably be expected to be achieved and the pessimistic duration is the longest duration likely under normal circumstances.

4.9.9.4   There are many ways in which the distribution of these values may be attributed to an activity, ranging from linear to parabolic distribution – many tools offer a dozen or more options. However, a starting assumption for each project activity may often be a 'triangular distribution'; the distribution is described as triangular because that is the shape of the probability graph for the activity duration.

4.9.9.5   For example, in Figure 4 the graph shows the triangular distribution for an activity with a minimum duration of 8, most likely of 14, and a maximum duration of 24.

4.9.9.6   From Figure 4 it can be observed that there is approximately an 18% chance of the duration being 14 days, approximately an 11% chance of the duration being 17 days and there is a 0% chance of the duration being either less than 8 or more than 24 days' duration.

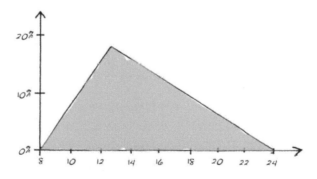

Figure 4 Triangular distribution of duration risk.

4.9.9.7   A description of the above distribution may be expressed as: 'the activity is likely to have a duration of 14 days. There is nil chance of it taking more than 24 days, and nil chance of it taking fewer than 8 days. The activity will probably be somewhere between 12 and 16 days in duration'.

---

[13] Where historical data is available, this will allow both the range and the distribution to be deduced. In the absence of reliable data, estimation is required.

4.9.9.8   Using the range information to apply a contingency to the activity duration will result in an expected duration of approximately 16 days, consisting of an activity of 14 days' duration and a contingency period of 2 days following it. Because this is calculated formulaically rather than by reference to specific risks, a contingency at this level must be considered to be a notional contingency to accommodate all risks of both parties and clearly identified in the planning method statement.

Contingencies at Medium Density

4.9.9.9   There is little scope at this level for notional formulaic calculations to accommodate unknown and unquantifiable risks.

4.9.9.10   At this density, contingencies must be clearly allocated to one party or another. There must be no contingency that is unallocated to an owner.

4.9.9.11   At Medium Density the risks should be clearly identified and a rational explanation set down in the planning method statement of the manner in which the possibility of the risk maturing has been allowed.

4.9.9.12   Allowances should be included in the schedule at this level for the assumed or allowed time needed for the completion of such unknowns as prime cost and provisional sums, the time effect of approximate quantities, and for the time to be taken up in completing the work associated with the expenditure of other financial contingency sums.

Contingencies at High Density

4.9.9.13   Simply because the predictive period is brief, the risks that need to be accounted for at High Density will tend to be significantly fewer than at other densities.

4.9.9.14   At this density, contingencies must be clearly allocated to one party or another and identified as such in the planning method statement. There must be no contingency that is unallocated to an owner and none that are not clearly justified in the planning method statement.

4.9.9.15   There may legitimately be risks such as adverse weather, unforeseeable ground conditions, utilities and third-party projects, plant breakdown, rework or absenteeism that may need to be allowed for at this density, but there should not be the need for design risk contingencies or implied variations at this stage.

4.10   Scheduling techniques

4.10.1   The range of scheduling techniques available and their application for use in an effective time model will largely depend upon the complexity and nature of the project, the nature of the work involved,[14] and the type of reporting required.

4.10.2   Critical path method (CPM) network diagrams are essential for illustrating the planned sequences, interrelationships and dependencies between activities, and for demonstrating the logic of carrying out the works. CPM networks also provide the necessary model for predicting consequences and managing the effects of intervening events by executing 'what-if' scenarios in the light of the intended logic. The robustness of the activity network is essential to subsequent analysis of the schedule. The principal alternatives are:

■   arrow diagram

■   precedence diagram

Part 4 - Developing the dynamic time model

---

[14] For example a 'Time chainage' chart may be more appropriate for a linear project such as a pipeline.

4.10.3   Depending upon the software product, the schedule may be configured to appear as one or more of the following:

■   an activity-on-arrow (ADM) diagram

■   a precedence (PDM) diagram

■   a linked bar chart

■   a data table

4.10.4   Because its content is most easily assimilated, for most Low Density reporting processes a bar chart is likely to be the preferred display technique. A linked bar chart is also easy to understand and, depending upon the ability of the software product to vary the visual scale of the image, may provide a useful figure for logic tracing. However, a network diagram based on logic is usually much easier to follow than a bar chart, because it is a horizontal representation of the work. The difficulty in reading a large bar chart with many lines of data is caused by the difficulty in tracing logic vertically. For Medium Density and High Density scheduling logic tracing, an ADM or PDM diagram is usually necessary.

4.10.5   A data table will be necessary for productivity analysis and is usually essential for quality assurance and auditing of a CPM network.

4.10.6   Effective time management cannot be achieved unless the planned sequence of activities in the working schedule is monitored and updated to reflect actual progress and the effect of intervening events, and the consequences are controlled. This process cannot be managed effectively if the planned schedule is based upon a simple bar chart of activities. Consequently, a logic linked activity network is the foundation upon which effective schedules and project controls are built.

4.10.7   Whichever scheduling technique is adopted, it must be anticipated that change will occur during the project timescale and that adjustments and amendments will need to be reflected within the time model.

4.10.8   There are various ways of illustrating scheduling techniques, some of which are as follows:

■   bar chart

■   line-of-balance diagram

■   time chainage diagram

■   arrow diagram

■   precedence diagram

■   linked bar chart

Bar charts

4.10.8.1   Bar charts have no logic. They are useful for illustrative purposes in high-level, Low Density reporting, but are of no value in the management of time in major projects.

Line-of-balance diagram

4.10.8.2   The line-of-balance diagram (see Figure 5), which is commonly used to indicate the progression of resources through a multiplicity of areas, is a useful technique for assessing the underlying philosophy of resource logic. The line-of-balance technique has been proven to be useful in the management of linear projects (e.g. rail, road and pipeline projects) and in repetitive modular work (e.g. structural floors of multi-storey buildings). In major projects it also has a useful role to play in planning and resource scheduling of the interface between different trades working in large areas, for example:

■ piling

■ ground beams

■ formwork

■ reinforcement

■ placing concrete

■ curing

■ strike formwork

■ backfill

or, in building works with services above suspended ceilings:

■ ductwork

■ first fix electrical

■ first fix plumbing

■ first fix security and communications

■ suspended ceiling grid

■ ceiling tiles and second fix services and fittings.

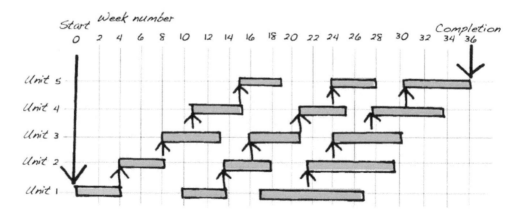

Figure 5 Typical line-of-balance diagram.

Time chainage diagram

4.10.8.3   Time chainage diagrams are commonly used in connection with work of a linear nature (e.g. road, rail, tunnelling and pipeline operations) to provide a graphical indication of both time and location of each work front (see Figure 6). Time chainage diagrams have limited application in the construction of major building projects, but can be very effective on engineering projects.

4.10.8.4   Owing to the complex nature of a time chainage diagram where multiple trades and work fronts may be included on a single diagram, the final output can become visually confusing and difficult to understand.

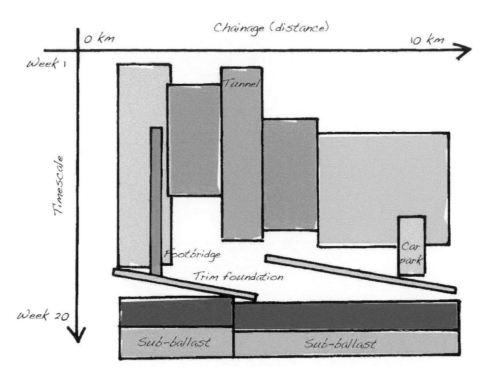

Figure 6 Typical time chainage diagram.

4.10.8.5   An example of an ADM network identifying the activities on the network is provided in Figure 7.

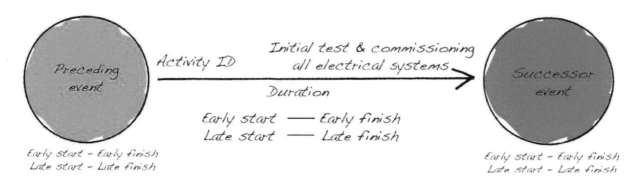

Figure 7 Typical activity diagram showing node–activity relationships.

Arrow diagram method (ADM)

4.10.8.6   This is the original critical path network technique developed to utilise the power of mainframe computers in the early 1960s. Before the advent of scheduling software packages for personal computers, in order to avoid the cost of having a mainframe computer, it was common to develop a schedule activity network by hand, where the activity was designated by the arrow and the logical interface of activities was represented by a node identifying the interface, as in Figure 8. The activity descriptions were originally contained in a separate document in order to reduce the physical size of the network.

4.10.8.7   In Figure 8, the 'dummy' illustrates a dependency between the completion of Activity C and the commencement of Activity B, which is not represented by any work activity. A similar technique must be adopted for lags.

4.10.8.8   ADM networks are less prone to manipulation than PDM networks.

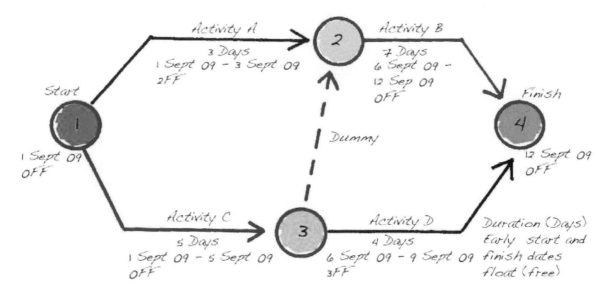

Figure 8 A four-activity ADM network.

Precedence diagram method (PDM)

4.10.8.9   This is the scheduling method adopted by most modern scheduling software packages. In this technique, the activity is designated by a node containing the activity-related information, and the logical interface of activities is represented by an arrow identifying the dependencies. An example of a typical PDM node is illustrated in Figure 9.

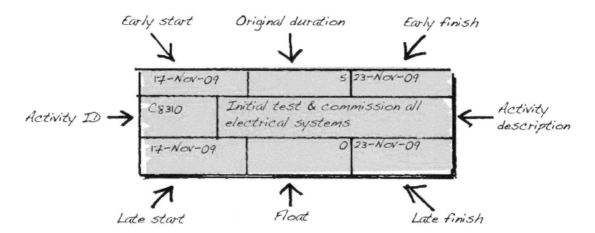

Figure 9 A precedence diagram method node.

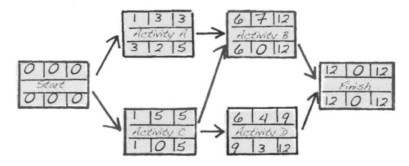

Figure 10 A four-activity PDM network.

4.10.8.10   In Figure 9, the activity is identified by an activity description and activity ID; in this figure the dates are illustrated as calendar dates and the duration and float identified in weeks, days or hours.

4.10.8.11   When linked together a simple precedence diagram looks like that illustrated in Figure 10.

Linked bar chart

4.10.8.12   This is another scheduling method offered by most modern scheduling software packages. In this technique, the node is drawn as a bar, against a time scale in which the length of the bar is proportionate to its duration, and with its start and finish dates aligned with a calendar at the top and/or bottom of the figure. The activity-related information is contained in columns of data to the left of, and in line with, the activity bar. The logical interdependency of activities is represented by an arrow identifying the interface. An example of a typical linked bar chart, using the same data as is used in the PDM network in Figure 10, is illustrated in Figure 11.

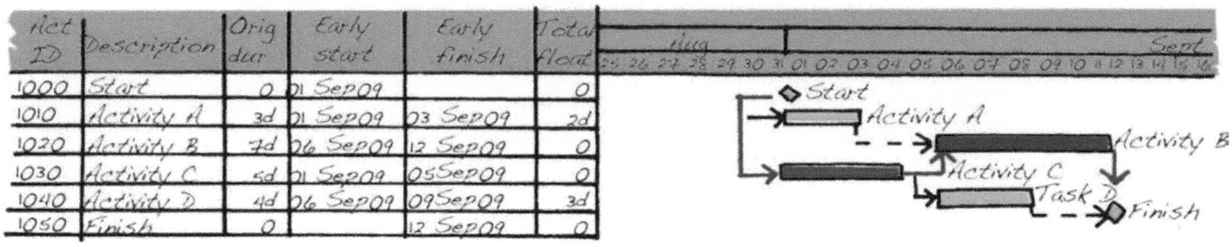

Figure 11 A simple linked bar-chart network.

4.10.8.13   For most purposes, because of its infinite flexibility in report content and its ease of interpretation, this is the network illustration method of choice for project reporting. However, as a scheduling method it has shortcomings in that:

- the construction of a network as a linked bar chart tends to encourage the user to think in terms of lists and dates rather than in terms of logic and sequence;

- the durations (represented as bars) tend to encourage the user to move the bars to 'paint the picture' which the user wishes to dictate, rather than to permit the software to calculate the sequence and dates; and

- it tends to be more difficult, than with other scheduling methods, to make changes to the logic with the addition and/or deletion of activities and even to make changes of logic and sequence to the existing activities.

Building information modelling

4.10.8.14   The 4D model should be continually developed as additional design and time-related information become available. These updates can be synchronised with the 4D model to reduce rework, but additional items and large changes will require special attention.

4.10.8.15   Most 4D modelling software packages synchronise with common scheduling software, but care should be taken to ensure the 4D modelling software interfaces with the selected scheduling software to produce the time model.

4.10.8.16   Colours, fonts and graphics may be used to assist in identifying areas of work, or works in progress or in delay.

4.11 Work breakdown structure

4.11.1   A work breakdown structure (WBS)[15] is a means of breaking down the full project scope of work into manageable component parts by reference to a structured hierarchy.

4.11.2   The establishment of a WBS assists with control and visibility of project components and makes them into manageable elements of work. Care needs to be exercised in deciding the level of detail in the WBS and the relationship between WBS elements and the activities in the schedule. If the WBS is carried through to the High Density level, each element will be assigned to specific resources.

4.11.3   The WBS must provide a hierarchical structure, starting at its highest level with the project as a whole, and at the lowest level ultimately identifying all work to be accomplished, for example, project, phases, areas, work packages and deliverables.

4.11.4   Each project will have its own idiosyncrasies and will necessitate the design of a workable template for the assignment of activities and deliverables according to the selected WBS structure. Where a 4D model is to be used in a BIM environment, the 3D model should be built with the 4D model in mind. For ease of updating, the WBS used in the model can be used to guide the layers in the 3D model for the most efficient 4D assignment and synchronisation.

4.11.5   The first step in the preparation of a schedule is to define and implement a work breakdown structure. A simple project WBS based upon zones of activity and stages of work is demonstrated in Figure 12.

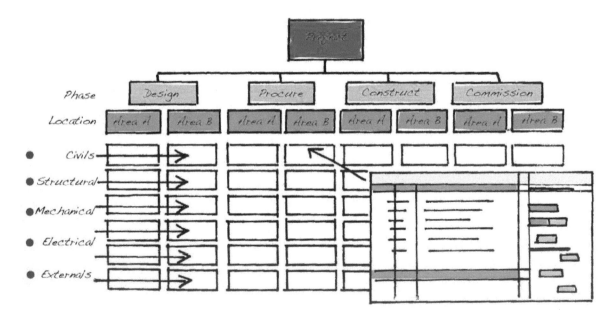

Figure 12 An example of a functional project WBS.

4.11.6   An alternative WBS hierarchy illustrating a relationship of resources is illustrated in Figure 13.

4.11.7   Although the WBS may be amended at a later stage in the project, the resultant change may cause confusion with interrelated and associated controls

---

[15] For a definitive explanation of WBS, see US Military Handbook 881C; Australian Standard AS4817: 2006; and BS6079 – 1: 2002.

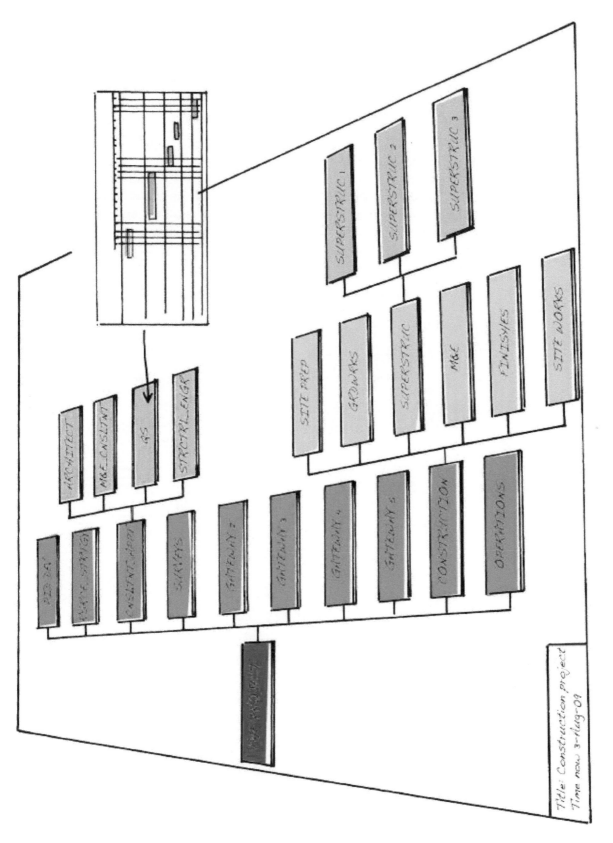

Figure 13 A typical WBS showing work-packages/professions/trades.

established against the same structure, for example, cost systems, progress measurements and so on.

4.11.8   Therefore, a clear, well-thought-out WBS must be prepared at the commencement of the project. When complete the WBS should result in a number of discrete work packages of 'deliverables' that in total define the full project work scope and encompass allocated, measurable periods of time and resources.

4.11.9   Once the structure of the WBS is established, a range of project controls can be assigned to each of the work packages. These controls typically include the assignation of a responsible person for the delivery of the work package, key dates to be met within that package, a budget and an indication of key deliverables.

4.11.10   The WBS may also be integrated with other organisational structures such as the cost breakdown structure or the organisation breakdown structure (OBS) in a matrix fashion, to provide an alignment of work packages with associated budgets/costs, or organisation responsibilities. Care should be taken, however, in the development of such a framework, as it is all too easy to confuse the straightforward intent of this tool (which is to provide clarity of work, deliverability, timing and control) by introducing an organisational structure which is too complex or unwieldy.

4.11.11   An integrated WBS, CBS and OBS is shown in Figure 14.

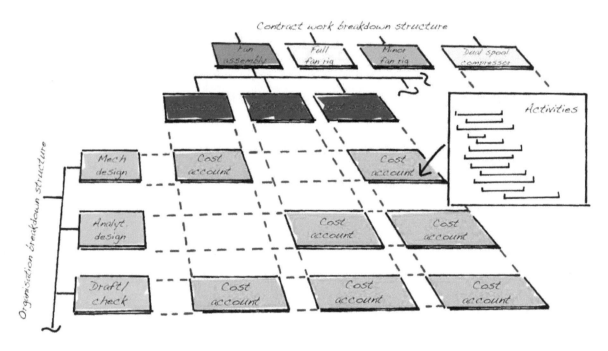

Figure 14 An integrated WBS, CBS and OBS.

4.12   Schedule communication

4.12.1   In major projects it is impracticable to use the whole of the schedule at any one time due to the number of activities involved. For effective communication the schedule should be designed to allow for segmentation and summarisation to different densities for differing purposes (see Figure 15). Most project scheduling software packages facilitate this hierarchical structuring by virtue of a summarisation or roll-up facility.

4.12.2   Typically, standardised layouts must be prepared to illustrate such features of the works as:

■   the intended sequence and timing of specific trades or processes;

■   information release dates, submission and approval dates;

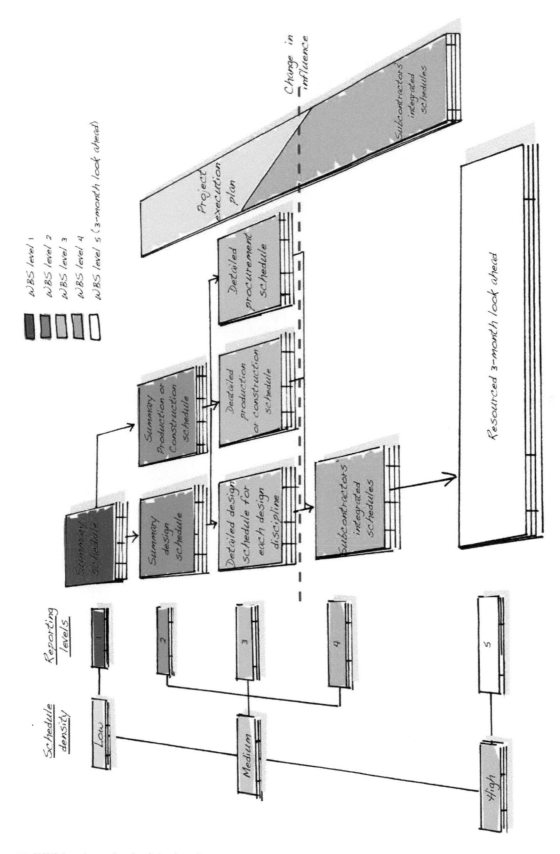

Figure 15 WBS levels and schedule density.

■ the effect of progress achieved from time to time on each completion, sectional completion or key date; and

■ the effect of any one or more intervening events on each completion, sectional completion or key date.

4.12.3   Most scheduling software packages will offer multiple formatting options. The temptation to use as many of these combinations as possible in any one chart should be avoided as the result can distract the reader from the information. Good graphic design is required to effectively communicate schedule information.

4.12.4   However, for clarity, care should be taken in the selection of fonts and colours so as to distinguish clearly between the structural elements of the schedule. Cartographers know that a map of any degree of complexity can be filled with only four colours without planes on a common border having the same colour. The same degree of clarity can be achieved on a schedule by limiting the number of colours used and avoiding multiple patterns.

4.12.5   In the selection of colours and patterns it should be borne in mind that, for some purposes, diagrams may need to be reproduced or transmitted in black and white.

4.12.6   The following provides an outline of the usual WBS-related reporting levels that the design of the schedule should allow for at the outset.

Executive summary report

WBS Level 1 report
4.12.6.1   This represents major milestones in the schedule; it highlights major project activities, milestones and key deliverables for the whole project. It is used to summarise the schedule in Low Density reports and other documents when a more detailed schedule is not required and may reasonably be prepared as a bar chart.

Senior management report

WBS Level 2 report
4.12.6.2   Maintained as a summary of the WBS level 3 schedule, it depicts the overall project broken down into its major components by area and is used for Low Density management reporting.

Project manager's report

WBS Level 3 report
4.12.6.3   Maintained as a summary of the WBS level 4 schedule for reporting status to senior management and to report monthly status to the employer. It is usually in Medium Density.

Section manager's report

WBS Level 4 report
4.12.6.4   WBS level 4 is the detailed working level schedules, where each schedule is an expansion of part of a WBS level 3 schedule and is established within the integrated project schedule. This is the working level schedule displaying the activities to be accomplished by the project workforce in Medium Density. The dates generated by these activities represent the anticipated start and completion of work required to complete the project.

Short-term look-ahead report

WBS Level 5 report

4.12.6.5   This illustrates the further breakdown of the activities of a WBS level 4 schedule into a High Density, short-term schedule used to map out the detailed activities needed to coordinate day-to-day work by resources allocated to specific areas of the project.

4.13   Calendars

4.13.1   Calendars have to be defined in order to establish the amount of working and non-working time against which an elapsed activity duration can be calculated.

4.13.2   Typically, calendars are established for the time generally available to the project, taking account of the working week, weekends, holiday periods and the like; special calendars may be required for specific resource working hours (e.g. a resource may only be available to the project for certain days of the week, or restricted to certain hours of each day, or months of the year). It is also possible to establish calendars for exceptional periods of time (e.g. 24-hour working).

4.13.3   Most modern scheduling software comes with a selection of pre-set calendars which may often be used as a starting point from which to develop project specific calendars.

4.13.4   There are a number of ways in which the effect of calendars may be illustrated on the network, but whichever method is selected, the chosen combination of display facilities should produce clarity.

4.13.5   For example, on a linked bar chart, the effect of calendars can be illustrated so that:

■   non-working periods are illustrated as coloured vertical bands, in effect blanking the non-working days;

■   selectively, illustration of some non-working periods, such as weekends, can be omitted in order to reduce visual clutter;

■   the activity bars can be shown in the foreground, or background, in relation to the non-working bands.

4.13.6   Some scheduling software will also permit the allocation of work stoppage periods (sometimes known as exceptions) to which different colours and patterns may be applied. Again, the use of colours and patterns should be used only to achieve clarity.

4.13.7   The project calendar will set out:

■   the normal working day, and

■   working time within working days to accommodate the number and timing of hours per shift and shifts per day.

4.13.8   The time unit used in planning and scheduling must be the same as the time unit against which production records can be kept.

4.13.9   For most building work, a time increment of a day is usually considered sufficient to be meaningful and anything more detailed is likely to be unmanageable.

4.13.10   However, some types of work, particularly those concerning facilities in use, may require short possessions to be planned by the hour and minute. On larger engineering projects, time units of weeks or months may be appropriate time units. Where relevant, the schedule design must permit the use of appropriate time units.

4.13.11   The calendar will also identify exclusions from the normal working day to accommodate, for example:

■ weekends

■ statutory holidays

■ religious holidays

■ other industry-recognised holidays

■ local holidays, etc.

4.13.12   Separate working periods may need to be defined to take account of the working periods of, for example:

■ specialist contractors

■ subcontractors

■ suppliers

4.13.13   For any resource, working periods may also need to be defined to take account of such things as, for example:

■ limited possessions

■ weather cycles to which the resource is sensitive

■ planned overtime

■ power outages

■ environmental considerations[16]

■ earthworks in winter

■ temporary traffic diversions

■ plant maintenance downtime

4.13.14   Calendars should be set up to account for anticipated non-working periods over and above weekends and public holidays (such as railway possessions and power outages), embargo periods (for such things as earthworks in winter, river crossings in the fish breeding season), road closures, planned plant maintenance downtime, and the like.

4.13.15   However, because of the difficulty in identification and traceability, calendars must not be used to allow for contingency periods by identifying, for example, what would otherwise be working days as notional non-working days to allow for potentially adverse weather conditions. The correct way to make such allocations is as contingency activity periods which can plainly be seen and audited.

4.13.16   Calendars can also be required for activities that are not related to the site operating calendar such as:

■ submissions and approval consideration periods;

■ procurement and delivery periods;

■ mobilisation periods; and

■ concrete curing time.

---

[16] For example, the effect on protected species and breeding seasons.

4.13.17   Depending upon their purpose and any conditions defined in the contract, such calendars may or may not recognise religious, industry-recognised and public holidays, and weekends, and each must be considered on its merits. For example, concrete cures 24 hours per day, 7 days per week, whereas a client approval period may be specified in terms of a standard five-day working week.

4.13.18   Project scheduling software products do not all deal with multiple calendars in the same manner. Accordingly, it is strongly advised that calendar allocations are kept as simple as possible in order to gain a meaningful understanding of their effect on time. It should be borne in mind that the effect of the allocation of multiple calendars will be dictated by the scheduling algorithm used by the scheduling software, and for any given combination, different software may produce different effects on the calculation of criticality. Accordingly, it is important that the scheduler and the management team understand how the chosen software uses calendars in relation to both activities and resources.

4.13.19   Resource calendars are usually established by specific, named calendar definitions, or by defining the actual date changes in each particular specified resource availability. These may be allocated to the resource in the resource definition, but the way calendars can be allocated to resources and their effect on the scheduling process is very much dependent upon the characteristics of the software being used.

4.13.20   Resource calendars will usually override the default activity calendar when rescheduling, so care should be taken to allocate non-working periods correctly when identifying the working time of a particular resource.

4.13.21   In addition to identifying working days, the calendars also identify working hours within the working days. These are referred to as 'work patterns'. Work patterns are the series of working and non-working times in any single working day. Only one work pattern can be assigned to any single working day in a single calendar, but different calendars may have different work patterns on the same day. At the extremes, some scheduling software will allow the work pattern time to be allocated in minutes and seconds; others may be coarser and limit the options to nothing more precise than half days. Depending upon the scheduling software, work patterns may be allocated to apply to a year at a time, while in others the options may be limited to weeks or months at a time.

4.13.22   Although it is technically possible to incorporate a night shift into what is otherwise a day-shift work pattern, in some software products some facilities may be adversely affected. Generally if a night shift is needed, then a different calendar should be allocated for it.

4.13.23   Different work patterns can be assigned to particular seasons in the same calendar, for example:

■   longer work periods during daylight saving and shorter work days during inhospitable winter working conditions;

■   minimal work periods to allow for extreme weather conditions in harsh environments where working in the middle of the day is either the only possibility or, alternatively, impracticable.

4.13.24   Work patterns must not be used to identify contingency periods by identifying, for example, what would otherwise be working hours as notional non-working hours to allow for potential lost productivity. The correct way to make such allocations is as contingency activity periods which can plainly be seen and audited from time to time.

4.13.25 The possibility of increasing working hours beyond the norm in order to shorten durations should be avoided. Increased working hours for other than a brief period will always result in a loss of productivity, and unless productivity output and increased risk of rework is also taken into consideration, the apparent shortening of the duration is likely to be misleading. Care should be taken to ensure that the allocated working patterns are consistent with local health and safety requirements.

Calendars at Low Density

4.13.26 Initially, calendars may be limited to those identifying the activity-related working week, together with statutory and other holidays.

Calendars at Medium Density

4.13.27 At this level all relevant activity-related calendars and the major resource-related calendars should be in place.

Calendars at High Density

4.13.28 By the time the work is about to start on site, all resource calendars must be established and allocated to the appropriate resources to be employed.

4.14 Work type definition

4.14.1 The work types which should always be identified are:

- key dates, sectional completion dates and the completion date
- final dates for receipt of design information
- contractor's design and approval
- statutory approvals, permissions and works
- sample production and approval
- off-site fabrication
- equipment order, procurement period, delivery and installation
- access and egress to and from the site, and parts of the site
- temporary works
- temporary plant order, installation, commissioning and decommissioning
- construction
- fixture and fittings
- testing and commissioning of installations, and
- handover process.

4.15 Activity identifier coding

4.15.1 An activity identifier code (ID code) is the unique unifying data identification that ties together the schedule, planning method statement, progress records and other project-related data. It is important that the structure for the naming of activity ID codes is defined.

4.15.2   This usually involves determining a coding structure that is sufficiently flexible to provide identification by reference to a variety of criteria. For example, the activity content code could be based upon:

■   the project and subproject zonal configuration;

■   the nature of the work to be scheduled;

■   the contractual significance of the work.

4.15.3   Some software products will automatically suggest an activity identifying code, and/or increment it according to selective data set up in the project definitions in many software tools; the activity ID is one of several code fields. In these circumstances, consideration should be given as to how the different fields will be used in combination.[17]

4.15.4   The activity ID should be capable of being broken down into related sub-activities, from Low Density to Medium Density and to High Density, as the work progresses, without loss of an audit trail through the development of the activity detail. So, taking the same example as for description, the activity IDs could be constructed as illustrated in Figure 16.

4.15.4.1   The importance here is not the complexity or simplicity of the coding formula, but that it creates a meaningful activity ID which is suitable for its purpose in the circumstances and can be broken down into detail as the schedule density requires.

4.15.5   In Figure 16, the descriptions have been kept simple for clarity, but see Figure 17 for a breakdown of a more detailed approach to activity descriptions.

4.15.6   In this example, the letter Z has been reserved for a code-identifier column which has no data and, where there are no more than 25 buildings and no more than 25 zones, a single letter has been used to identify each.

Activity ID at Low Density

4.15.6.1   All the work to be coded occurs in Building A and is in Zone B of Building A; therefore, no matter at what degree of density, all activity IDs commence with the two letters AB.

4.15.6.2   At Low Density, all we have is a single activity called 'substructures', which in this example is called Area A. Accordingly, the activity commences ABA.

Activity ID at Medium Density

4.15.6.3   At Medium Density, the 'substructures' are broken down into 'excavations', 'piling', 'ground beams' and 'floor slabs', each of which is given an identifier letter so that in this example 'ground beams' is called Section C. Accordingly, the activity commences ABAC.

Activity ID at High Density

4.15.6.4   At High Density, the 'ground beams' are broken down into 'formwork', 'reinforcement', 'placing concrete', 'curing', 'strike formwork' and 'backfill', each of which is given an identifier letter so that in this example 'reinforcement' is called Item B. Accordingly, the activity commences ABACB.

---

[17] See 'Activity content codes' at Section 4.17 and 'Activity cost codes' at Section 4.18.

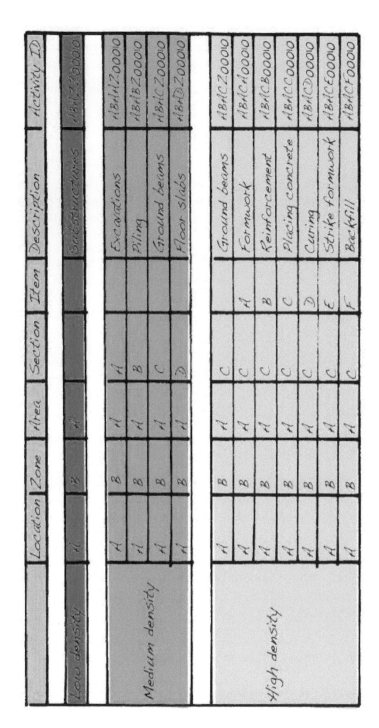

Figure 16 Example of activity ID coding structure.

Activity ID trailing numbers

4.15.6.5   The trailing numbers start at 00010 and would normally progress in tens to leave room for any other activities needed to be inserted at a later date within the bracket of 10 (so as not to break the numerical sequence). Thus, the next breakdown of the 'reinforcement' activities might be 00020, 00030 and so on, and if at a later date another reinforcement activity were to be needed between 00020 and 00030, it would have the activity ID of ABACB00025.

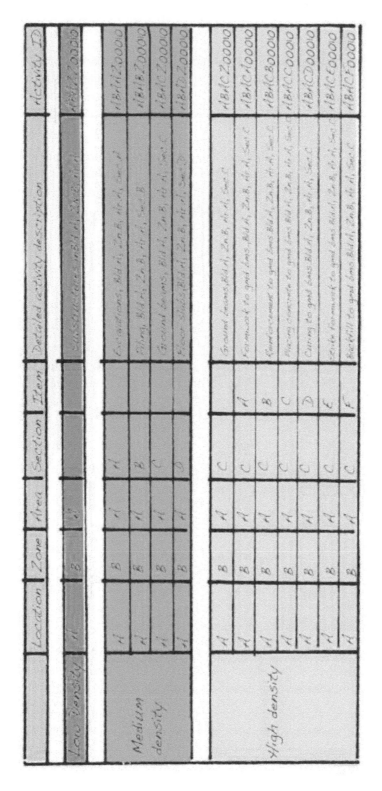

Figure 17 Unique activity descriptions.

Simplified Activity ID

4.15.6.6   Alternatively, a simpler form of activity ID coding may be adopted in some circumstances, whereby in the development schedule an activity ID is assigned depicting only the phase and zone of operation, for example, DA1000 = design, area A, activity 1000. This may then be developed into a working schedule by adding further breakdown as DA1000.01 to 99 and so on.

4.16   Activity description

4.16.1   Activity descriptions capture the essence of the intended project deliverables. Given that they are often truncated summaries of the fuller descriptions contained elsewhere, for example, in the employer's requirements, it is important that whatever density of schedule is adopted, the activity description is clear and unambiguous.

4.16.2   Most software products have a limited field for the activity description. Thus, apart from configuring a unique identifying code for each activity, it is important to establish a convention for the naming of each activity so that the descriptive content of each is unique and unambiguous.

4.16.3   Depending upon the purpose of the schedule, at feasibility stage the activity description may be so coarse as to facilitate no more than an intended duration, from start to finish, of each building within a group. On the other hand, at construction level, significantly greater detail will be required for time management. As the density of the schedule increases, so the clarity and unique quality of the description becomes more important.

4.16.4   The activity description must clearly describe the scope of the works and should be phrased with a verb-noun combination (e.g. 'excavate trench', 'pour concrete', 'erect steelwork') to which more precise descriptive elements can be added as the schedule density increases (see below).

4.16.5   Milestone descriptions should relate to an event or deliverable finishing or starting, with the verb-noun combination rule still being applied e.g. 'start excavation'.

4.16.6   The activity description must be clear, concise and fit for its purpose at each level of density.

Descriptions at Low Density

4.16.6.1   Initially, the schedule may be generated at a time in the project when the scope of work is not fully defined. Accordingly, for Low Density schedules an activity description that facilitates later subdivision will be required (e.g. substructures).

Descriptions at Medium Density

4.16.6.2   It is important to note that the early construction-related schedule will be submitted to fulfil contractual obligations and it is of paramount importance, therefore, that all elements of the work are adequately described.

4.16.6.3   It is worth keeping the activity description at the early stage of development at a reasonably high level in order to maintain a degree of scheduling flexibility once the scope of work is further defined and detailed (e.g. excavations, piling, ground beams, floor slabs).

Descriptions at High Density

4.16.6.4   The activity description must clearly identify, unambiguously, exactly what work is to be done and where it is to be carried out. Without such clarity, it will be impossible to accurately record the progress of it, for example:

■   ground beams

■   formwork

■   reinforcement

- placing concrete

- curing

- strike formwork

- backfill

4.16.6.5   An example of how the increasing density of activity description works in practice is given in Figure 17, which, by using the same example data as in Figure 16, illustrates how the descriptions are nested but rendered unique by reference to the data identifying such things as Location, Zone, Area, Section and Item.

4.17   Activity content codes

4.17.1   Except in unusual circumstances, the schedule will be reviewed and revised from time to time. Intervening events will occur which will need to be taken into account in the forward planning of the work, and progress achieved will be different from the productivity expected. In order to manage the time implications of these departures, an activity coding structure which can usefully assist in filtering operations is important where a single view of individual trades or disciplines is required. When designing the content code structure, consideration should be given to such matters as:

- changes to schedule density;

- change of calendar;

- resource allocations;

- creating or amending hammocks;

- inserting resource-driven links; and

- providing a short-term schedule of work for each trade or discipline.

4.17.2   Different scheduling software will deal with activity content coding in different ways, and it is important to understand how the scheduling software handles activity codes before attempting to apply them.

4.17.3   Activity content codes should identify the various attributes of the schedule as fields, the values of which will facilitate organisational changes, and facilitate filtering of important parts of the schedule. Content coding is thus used to provide:

- visual impact and identification of an activity group on the schedule; and

- an aid to filtering, linking and organising activities into suitable 'views' for quality assurance, auditing and reporting purposes.

4.17.4   Most current scheduling software products will offer the scheduler a multiple choice of formatting options for the presentation of assigned activity content codes. In so far as the software defaults to using particular colours or patterns to indicate critical and non-critical activities, they must not be used for the indication of other fields, or values in other fields.

4.17.5   If the scheduling software supports the allocation of multiple tasks to a single activity for repetitive projects such as high-rise structures, road operations, pipelines, or procurement stages, then the judicious use of activity content codes can assist in the presentation of activities up or down the project, visually similar to the picture generated by a time chainage schedule.

4.17.6   Project scheduling software is based upon a database application containing the facility for the creation of many different data fields and values within those fields. One type of field is known as the 'activity content code'. By the attribution of codes and values to a group of activities, they can be searched, filtered and displayed as a discrete group to facilitate auditing and quality assurance, review, revision and reporting.

4.17.7   Typically, the schedule may be organised either by means of the WBS or by the use of content codes. However, in so far as the WBS is a relatively static structure, because they are attributed for particular purposes, activity content coding provides an infinitely variable structure against which the schedule may be rearranged.

4.17.8   Examples of typical descriptive content codes and values that may be attributed are illustrated in Figure 18.

Descriptive Fields	Descriptive Values
Location	Building 1, Building 2, River crossing
Area	Basement, Ground, 1st Floor, 2nd Floor
Zone	Gridlines, A1-G5, GridlinesA6-G10, Gridlines A1-G15
Department	Design, Purchasing, Information systems, Construction
Responsibility	Project director-Alan Cappins, Chief architect-Dave Schmit, Director of purchasing-Ellen Ulrich
Phase	Conceptual design, submittals and approvals, shop drawings, procurement, construction, testing
Section	Foundations, structural frame, cladding, installations
Events	A1, A12, CVC4, Claim1, Claim2, Claim3

Figure 18 Example of descriptive activity-content codes and values.

4.17.9   It is preferable to set up the activity content code groups that are most likely to be applicable to the schedule, in advance of setting down any activities to which the fields may be applied (see Section 4.16). Due account should be taken of 'common' code groups that may be applicable to an organisation or to a project, such that duplication of code groups is avoided within other areas of the enterprise.

4.17.10   Activity content codes and values may be added at any time during the planning process. However, once they have been attributed to activities, the subsequent amendment of the order in which the codes or values were created, or their structure, may cause significant difficulties and should be avoided.

4.18   Activity cost codes

4.18.1   Cost coding works in a similar way to activity content coding in that, given particular cost budgets or fields, values can be attributed to them in relation to particular resources or activities.

4.18.2   An effective cost coding system is an integral aspect of project control. If adequately configured to do so, it can be used effectively in calculating interim costs from interim progress updates and can help to avoid potential misalignment of anticipated completion values.

4.18.3   Cost codes also facilitate an alignment between cost accounting and management accounting so that the financial reporting tends to be more accurate.

4.18.4   A cost coding system does not need to be complicated in order to produce valuable data, and should be designed to meet the complexity level of the other required project controls. The cost codes and structures should be considered in much the same manner as the establishment of the WBS and content codes. Their application should be thought through in advance of their application onto the schedule, such that the desired level of analysis and reporting can be achieved.

4.18.5   Cost codes are used in much the same manner as content codes and work breakdown structures, in that they are applied to the activities in a schedule for the effective subdivision of budgets and cost collation, so that each cost package may be monitored and controlled.

4.18.6   Cost codes arranged in a structure aligned to the WBS will offer significant advantages to the project team. For example, the cost manager can use the combined sets of information to predict the cost estimate to complete for any cost package based upon the level of progress, cash-flow forecast and productivity achieved. Most software packages facilitate reporting against this data to provide a dynamic and integrated view of project progress related to earned value against predicted cost.[18]

4.18.7   The application of costs to activities and resources facilitates the alignment of project controls between the time manager, scheduler and the cost manager. However, the degree of detail will largely depend upon the density of the schedule.

Cost coding at Low Density

4.18.7.1   At this level of density, cost data may reasonably be represented by major budgetary groups and/or predicted payment milestones in the schedule. Given this level of schedule density, it cannot reasonably be expected to contain more than a headline view of budgets and will normally contain a high level of contingency.

Cost coding at Medium Density

4.18.7.2   At this level, all relevant activities and many of the resources should be known. Cost coding structures at this level are more often assigned to project activities and the accuracy of their baseline values is of importance to out-turn cost predictability.

Cost coding at High Density

4.18.7.3   By the time the work is about to start on site, all resources must be known and the resources to be employed allocated to the activities on which the resources are to be expended. Cost coding structures at this level are more often assigned to resources, and the accuracy of their baseline values and progress cost updates is vital to the management of interim payments and competent project cost control. It is, of course, paramount that the accuracy of the progress and cost data entered into the schedule periodically is based upon contemporary evidence and is reflective of the resources actually employed.

---

[18] Earned Value Management (EVM) is a standardised approach to financial performance management required under some engineering and construction contracts; the foundation for an effective EVM system is a well-constructed schedule that is kept up to date. See Section 5.9.

4.19 Activity duration

4.19.1   One of the most important considerations in the development of the content of an activity network is whether the duration of any activity is to be determined inductively or deductively. In other words, the scheduler must determine whether the duration of any activity is to be determined:

■   empirically, by setting down the date it will start and finish, assuming that sufficient resources can be made available to perform to that standard; or

■   deductively, by calculating the duration from the quantity of work to be performed, the resources which reasonably can be allocated and their rate of performance.

4.19.2   Most scheduling software will offer alternative activity rules for dealing with duration. The default is usually to give priority to the allocated duration in preference to any duration deduced as a result of applied logic or resources, although normally this may be changed to give priority to either of the other data categories.

4.19.3   When resource allocation and designated productivity are identified as determining factors in the calculation of an activity duration, the duration will be calculated by reference to the given resource data.

4.19.4   Depending upon the density of the schedule, the purpose for which it is to be used and the information available, an activity duration can be derived from:

■   experience;

■   industry standards;

■   benchmarking;

■   comparison with other projects;

■   calculation from resources; and/or

■   specification.

4.19.5   At Medium and Low Density, when activity duration is identified as to be calculated by reference to the logic of the schedule, for example, by extending an activity beyond its given duration to be consistent with finish-to-finish logic, it is usually referred to as a 'stretched' or non-contiguous activity. Scheduling software may permit this selection according to particular activities, or only in relation to the schedule as a whole.

4.19.6   Estimating durations by experience

4.19.6.1   Estimating duration based on experience of similar work in past projects is likely to be essential in Low Density and, to some extent, in Medium Density schedules.

4.19.6.2   Construction trade professionals and managers at the workface will tend to know from their own experience how long it is likely to take to perform a particular activity in ordinary day-to-day working conditions with a given gang size. They are also likely to be able to give some guidance as to the likely effect that restrictive conditions may have on their productivity.

4.19.6.3   In the absence of objective standards to work to, the scheduler will have to fall back on the experience of those who are familiar with the type of work. However, one technique still available is to gauge a range of possible durations from alternative sources and then to apply these in the following manner:

- the optimistic duration – the shortest time in which the activity can reasonably be completed;

- the most likely duration – the completion time having the highest probability;

- the pessimistic duration – the longest time that an activity may reasonably require.

4.19.6.4   From a collation of the above, the following formula will give a weighted expected duration:

$$Expected\ activity\ duration = \frac{Optimistic + (4 \times Most\ likely) + Pessimistic}{6}$$

Estimating durations using industry standards

4.19.6.5   Resource and productivity data should obviously be as realistic as possible for the prevailing circumstances and work type. The data can be obtained from a number of sources[19] but, from wherever it is obtained, its source should be recorded in the planning method statement

Estimating durations using benchmarking

4.19.6.6   With repetitive operations, a work study should be carried out to establish the performance baseline. If there are 500 bedrooms to be fitted out, the time it will take to fit out one bedroom will be an extremely important consideration: any error will be multiplied 500 times. Therefore, it will be very important to establish the correct duration for the first one by carrying out a work study. Similar considerations may be applied to the structural lifts of a tower block, piling, pipeline being laid, and, indeed, on a micro-level to any repeatable activity such as fitting a door. Thus, checking the exercise by a benchmark study, in sample, may be an extremely valuable exercise, and especially so if changes are later made which incur time-related and disruption-related costs.

4.19.6.7   Where benchmarked productivity data is available from the production data of similar work previously executed by the same contractor, this should be taken in priority over any other data which is not project-specific.

Estimating activity duration by comparison with other projects

4.19.6.8   In the absence of more objective data from which to work, taking as a baseline the known time from other projects of like content is often the preferred method of allocating time in Low Density schedules. Where there is a high degree of speculative time allocation, allowances need to be made to compensate for the higher risk of error.

4.19.6.9   In Low Density and Medium Density schedules, it is prudent to engage with those specialist subcontractors who may be or have been appointed in order to incorporate their experience into the estimates.

4.19.6.10   However, using as-built data from other projects is a technique of limited application in Medium Density schedules in which the majority of design will have been completed, and hence a more project-specific method of estimating activity durations will be appropriate.

---

[19] See Appendix 7 for a schedule of currently available sources.

4.19.6.11   The allocation of durations by reference to other projects of like kind has no application in High Density, short-term scheduling.

Calculating activity duration from resources and work content

4.19.6.12   The relevant resources are

■   labour

■   plant

■   money

■   materials, and

■   space

4.19.6.13   However, the order of priority of importance attributed to these resources will differ from time to time according to the type of work from project to project. For example, machine type and numbers, as well as the production capability of the plant, will be the driving factors in earthworks, but in most construction projects, labour is likely to be the driving resource for most types of work.

4.19.6.14   Four methods of resource scheduling are usually available, and it is useful if the scheduling software is able to calculate the schedule using all of them. Of particular importance is the facility for scenario modelling. The four methods are:

■   deadline critical resource levelling (also called 'smoothing') where overloading is reduced by moving activities within the available float, if any, and if there is no available float, the process will show overloading of resources, but the project end date will not change;

■   resource critical resource levelling (also called 'levelling') where the process will show a change in the project end date, if necessary, to reflect the time required to complete the project with the allocated resources;

■   resource aggregation using the activities' early start and finish dates; and

■   resource aggregation using the activities' late start and finish dates.

4.19.6.15   In the High Density part of the schedule, durations should be calculated by reference to resources and productivity rates actually achieved or realistically anticipated.

4.19.6.16   At High Density, the duration of an activity is to be a function of the quantity of work and the productivity quotient relevant to the resource allocated. This method of calculating durations may also be available to some activities in Medium Density schedules, but in Low Density schedules there will often be insufficient data available to produce meaningful calculations.

Specified activity duration

4.19.6.17   Notional periods for provisional work, work relative to prime cost sums and other contingencies should be defined and specified by the project team for inclusion in Low Density schedules and in Medium Density schedules in which the previously unknown work still remains undetailed.

4.19.6.18   Schedule density will be the deciding factor as to what is an appropriate and reasonable method for determining the schedule durations.

Activity duration at Low Density

4.19.6.19   Durations will often be derived by assessment from the experience of those associated with the project.

4.19.6.20   Through experience it may be known that the erection of steelwork with a single gang of men can proceed at 'x' tonnes per hour, therefore 'y' tonnes of steel will be erected in 'y' divided by 'x' hours (e.g. 500 tonnes of steel (y) at 5 tonnes per hour (x) = 500 divided by 5 = 100 hours).

4.19.6.21   At this level much of the work may not have been fully specified, and hence the definition of durations by reference to empirical data, with a high degree of formulaic contingency, will often be appropriate.

Activity durations at Medium Density

4.19.6.22   At this level the work can be expected to have been designed in detail and a higher degree of science should be applied to calculating the reasonable durations of activities.

4.19.6.23   At Medium Density there should be no approximation based upon an expected duration formula.

4.19.6.24   Activity durations will ordinarily be derived from one or more of the following:

■   the specified standards of performance (e.g. curing times, specified provisional and contingency periods, and similar provisions);

■   calculation by reference to quantities, approximate quantities, notional resources and standard production rates;

■   assessment by experience;

■   specialist subcontractor and supplier production data; and/or

■   historical production data.

Activity durations at High Density

4.19.6.25   This density governs work intended to be carried out in the next three months, and all information concerning the designed works and resources should be available.

4.19.6.26   There should be no assessment by experience alone at this stage, although experience will play a large part in determining whether planned production rates and resources are reasonable.

4.19.6.27   Activity durations will ordinarily be derived from one or more of the following:

■   the specified standards of performance (e.g. curing times, specified provisional and contingency periods, and similar provisions);

■   calculation by reference to quantities, approximate quantities and production rates;

■   specialist subcontractor and supplier production data;

■   actual production data from similar work;

■   historical production data; and

■   benchmarking.

4.19.6.28   In connection with the effects on durations of likely productivity, the following should also be considered:

■   physical working conditions;

■   safety requirements and labour agreements that may be in effect;

■   site downtime;

■   seasonal weather fluctuations;

■   seasonally related activities; and

■   resources for seasonal work.

4.20   Resource scheduling

4.20.1   Resource scheduling is quite different from the technique of adding resources to a critical path network for the purposes of calculating activity durations.

4.20.2   Resources are labour, plant, money and materials. Space and time can also be viewed as resources. One or more of these will be more important than the others in planning different types of work. In most construction projects, labour will be the most common variable across most types of work, but in earthworks, for example, machine type and numbers as well as the production capability of the processing plant will be more relevant.

4.20.3   Although, in building construction, work on site can be represented as a series of activities of short duration in discrete areas, which are to a great extent subject to a logical flow (and thus suitable for critical path network analysis), time modelling of design services and other types of construction work (e.g. civil engineering, mechanical engineering and process plant works) will not necessarily be subject to the same type of time control.

4.20.4   For example, in planning design services and in major land clearance, or cut and fill of land profiles, work may continue over a long period and may usefully follow some sequence (in design work, concept design will be completed before detailed design starts, and in earthworks, land will be dug out before depressions are filled), but not necessarily a sequence which is cogently linked from beginning to end of the operation as a whole.

4.20.5   Commonly, in earthworks the process may usefully be mapped out in what is referred to as a time chainage diagram, which illustrates, in linear terms, where work is intended to be carried out. Although such figures may be logically linked to some extent, because they are drawn rather than calculated from a database, they do not commonly function as a time model, and other methods must usually be used to manage time and predict the consequences of change.

4.20.6   In the time management of design services and in the circumstances where activities can be carried out over a long period in a multiplicity of areas, in any order (subject perhaps to limited sequences in any particular area), the time model will commonly focus on the management of resources and productivity, instead of critical path sequences within the particular activity and critical path sequences between sections of the activity and the interface with the remainder of the works.

4.20.7   It follows that such time models are often subject to a high degree of revision to accommodate the preferential sequences that are ultimately adopted.

4.20.8   Where the use of this type of scheduling is deemed desirable for part of the overall project, consideration should be given as to how the schedule will be integrated into the overall project schedule to retain vertical and horizontal traceability.

4.20.9   Different scheduling software will deal with resources in different ways, and it is important to recognise what the software will do and what it will not do before deciding how resources are applied to the schedule.

4.20.10   Not all resources are difficult to obtain (for example, resources which are available in abundance locally, are unlikely to constrain performance). Where resources are freely available, they may usefully be planned to be available in the quantities needed to pace the work of the project. Other resources are not so readily available, or benefit from consistent levels of utilisation; the use of these critical resources must be balanced first, to avoid overloading scarce resources and to minimise variations in demand. Then the impact of this balancing on the other resources can be shown graphically by individual resource histograms.

4.20.11   Balancing the critical resources can be achieved using different approaches, either manually or by applying different criteria available in most scheduling software packages in a series of iterations, resulting in alternative performance schedules for consideration by the management team.

4.20.12   The resources that should be identified separately are those categorised as:

■   separate contractors

■   subcontractors and trade contractors

■   consultants

■   employer

■   third parties

■   suppliers

■   plant

4.20.13   Where the working schedule is conceived as a 4D model in a BIM environment, the 3D virtual model will often be useful in demonstrating work that can be carried out over a long period in a multiplicity of areas, in any particular order. This is particularly useful for visualising any modelling of resource levelling and to assist with continuity of performance. Key site-related resources and equipment, such as cranes or scaffolding, can also be included in the 3D model to assist in resource planning.

Resources at Low Density and Medium Density

4.20.13.1   At Low and Medium Density there will often be insufficient data available to make precise computations of durations by reference to productivity and resource alone. At these densities, it is quite often the case that activities have a mixture of resources, and other methods of estimation are necessary. For example, estimation by reference to previous projects, experience, standard outputs and so on, is acceptable.

Resources at High Density

4.20.13.2   Resource scheduling is a time-intensive exercise that requires the participation of all those involved in the work to achieve a realistic outcome. However, in order to demonstrate a reasonable, rational and valid High Density working schedule, the work must be resource scheduled.[20]

---

[20] See Appendix 5 for two case studies of the benefits of resource scheduling at High Density.

4.20.13.3   The allocation of resources to activities to develop the High Density part of a realistic working schedule is a basic practice of network-based construction management required by the Guide.

4.20.13.4   Resourcing the High Density schedule requires the actual resources available for the work (and their productivity rates) to be used to calculate the duration of each activity so that the work can reasonably be expected to be performed as scheduled. If the High Density part of the schedule is showing unacceptable slippage or other problems, actions need to be planned and time allowed to change the actual resource availability or capability, and/or the sequence of work, to overcome the identified issues; the High Density part of the schedule must always realistically reflect the actual capabilities of the available resources so that it remains realistic and achievable.

4.20.13.5   At High Density, the planned duration of an activity is a function of the quantity of work, the productivity quotient and the quantity of the resource type to be deployed, in the formula

$$Duration = \frac{Quantity\ of\ work}{Productivity\ quotient \times Quantity\ of\ resource}$$

4.20.13.6   The productivity quotient of a resource can vary according to the crew size adopted for the work, but the relationship between crew size and productivity is not linear. Reducing resources below optimum introduces inefficiencies, as does adding more resources above the optimum. Loss of efficiency needs to be balanced against space, cost, time and logistic considerations in order to determine the appropriate crew size for the work, and therefore the productivity quotient and duration of the activity.[21]

4.20.13.7   The productivity quotient of a particular resource can be determined in a number of ways:

■   the actual productivity rates achieved on the project;

■   published output rates;

■   historical data from other projects;

■   advice from specialists;

■   personal experience;

■   benchmarking.

4.20.13.8   Resourcing the schedule will either validate the assessment of the durations adopted at lower densities, or require modification to the schedule to accommodate the actual anticipated performance of work. In addition, a resourced schedule permits:

■   An understanding of work flow for gangs/trades. A steady work flow will deliver the most efficient plan.

■   Reassurance to other interested parties (internal or external) on the practicality of the time model.

■   An understanding of any challenging resource requirements, for example, high-peak requirements, or lack of continuity for trades or gangs.

---

[21] See Appendix 5 'Case studies in High Density scheduling contents' for the benefits of this in practice.

Strategic resource allocation

4.20.13.9   Wherever possible, trade teams should be scheduled to maximise continuity and to minimise alternating work and non-work periods. However, once the allocation of resources has been made and activity durations recalculated according to the productivity quotients, it may be found that the resultant schedule shows discontinuous trade activities.

4.20.13.10   The object of levelling is to even out requirements for labour and plant resources so that, wherever possible, peaks and troughs are removed, and the consequential loss of productivity is avoided. The purpose is to avoid arriving at a schedule which, for example, requires 100 men and 20 machines one week, none at all the following week, and 50 men and 10 machines the week after that.

4.20.13.11   Resource levelling is achieved by adjusting the timing and/or duration of activities in order to even out the resource requirements. Peaks and troughs in resource requirements can be avoided by:

■   delaying non-critical activities, using float, so the work is scheduled in a period of low demand; and/or

■   reducing the resource allocated to an activity, thereby extending its duration; and/or

■   if necessary, delaying critical activities and hence delaying key dates, sectional completion dates, the completion date and/or contravening a constraint.

4.20.13.12   The resources can be adjusted according to:

■   a resource-critical resource levelling procedure (or 'levelling') that may delay a completion date; or

■   a deadline-critical resource levelling procedure (or 'smoothing') that may still leave resource overloads in the schedule.

4.20.13.13

The process consists of two steps:

■   estimating the amount of resources needed by each schedule activity; and then

■   adjusting the schedule by using float values computed in the mathematical network analysis to schedule activities within the assumed availability of those resources and/or adjusting resource requirements on activities by selecting alternates or adjusting allocations and durations, to optimize the outcome;

■   using float values computed in the mathematical network analysis to schedule activities within the assumed availability of those resources.

Where practicable, trade-related line-of-balance diagrams can be used to help to illustrate how a succeeding trade's start could be delayed to achieve continuous performance.

4.20.13.14   Both these methods make use of what would otherwise be available float overriding early start and finish dates and rescheduling them to dates when the resources become available.

4.20.13.15   Resourced schedules can provide resource histograms and cumulative graphs to assist those who have difficulty in reading raw data to review, easily and rapidly, at least one resource or element of the schedule at a time.

4.20.13.16   Figure 19 illustrates different resources in different colours prior to rescheduling and levelling. By review of an illustration such as this, the scheduler can examine the resources' use of the trade, shown in grey/white above the dotted line and consider the practicality of achieving the levels shown.

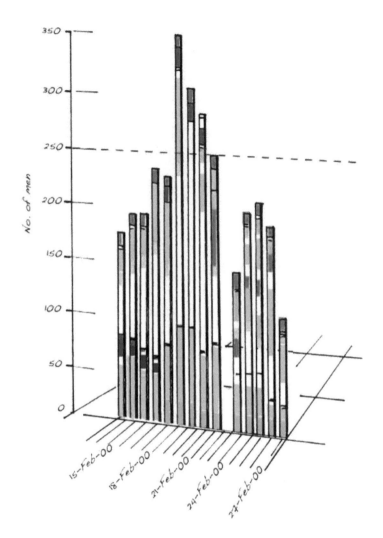

Figure 19 Multiple resources prior to levelling.

4.21   Permits and licences

4.21.1   The schedule should identify permissions which are required to be in place prior to work, or sections of work, being commenced or completed, and should identify separately those to be obtained by the employer from those to be obtained by the contractor.

4.21.2   For each permit or licence there will be a period of time to be allowed for:

- preparation of submission;

- decision period; and

- grant of permit or licence.

4.21.3   Depending upon whether the submission is likely to be contentious, it may also be advisable to allow periods for amendment of the submission and reconsideration (see also 'Risk and contingencies' at Section 4.9).

4.22 Utilities and third-party projects

4.22.1 Independent projects that have to be carried out at the same time as the works, and upon which parts of the works are often dependent, create special problems. For example, asset owners such as highway authorities and supply authorities for water, gas, electricity, communications and the like require special consideration in the design of a schedule, and in relation to each supply utility, provision should be made for the identification of:

- investigation of requirements;

- contract;

- grant of permit or licence;

- mobilization;

- work period; and

- connections and/or commissioning.

4.22.2 It should be borne in mind that work in connection with investigations may be dependent upon prior agreement for diversions.

4.23 Schedule logic

4.23.1 Unless there is a good reason to the contrary (in which case the reasons must be fully set out in the planning method statement), apart from the commencement milestone, every activity will have a link from a logical predecessor that connects to its start. Likewise, apart from the completion milestone, every activity should have a logical successor that connects from its finish. 'Open ends' (i.e. activities that do not have logical predecessors and successors) should be avoided.

4.23.2 Logic can be categorised in four ways:

- engineering logic

- preferential logic

- resource logic

- zonal logic

4.23.3 Some scheduling software products provide the facility for designating logic in the alternative, via an 'or' gate.

4.23.4 Whichever logic is adopted, the sequence of work incorporated in the schedule must mirror the planned intent for the future conduct of the work, from design through procurement and to work on site, at both the macro-level and at the individual activity level. The logical relationship between any two activities may require one to be completed before the next one starts, or may involve a planned overlapping of the work; the various possibilities are as follows.

Engineering logic

4.23.4.1 Engineering logic, sometimes called 'hard logic' or mandatory dependencies, is that which is indisputably essential to the process (e.g. foundations must precede superstructure, superstructure must precede fitting-out, and so on). This logic is immutable and generally cannot be changed.

Preferential logic

4.23.4.2   Preferential logic, sometimes called 'soft logic' or discretionary dependencies, is that which relates to managerial processes rather than engineering. For example, 'part A' of land clearance will be planned to precede 'part B' of land clearance, and 'part A' of the foundations will be planned to precede 'part B' of the foundations. Provided it is done in good time, so as to avoid unproductive resourcing, there is no practical inhibition to changing preferential logic to, in the same example, 'part B' preceding 'part A'. Changing preferential logic allows management to respond to changes in the project.

Resource logic

4.23.4.3   This is a variety of preferential logic whereby, for managerial reasons, particular resources are planned to proceed in a particular sequence. For example, a particular resource will commence in 'area A' and will be planned to proceed to 'area B'. Provided it is done in good time so as to avoid unproductive resourcing or any other loss of efficiency, there is no practical inhibition to changing resource logic to, in the same example, the resource commencing in 'area B' and proceeding to 'area A'). Using logic to control resource flows means management choose the optimum work sequence rather than relying on arbitrary decisions made by the software's resource scheduling algorithms.

Zonal logic

4.23.4.4   This is a variety of preferential logic in which, for managerial reasons, one part of the physical work is required to precede another. For example, work in the location of 'building A' will be planned to precede commencement of work in 'building B'. Provided that it is done in good time so as to avoid unproductive resourcing, there is no practical inhibition to changing zonal logic so that, in the same example, 'building B' is carried out before 'building A'.

4.24   Density logic

4.24.1   Regardless of density, in order that it function as an effective time management tool, the schedule should be fully linked. However, it is important to ensure linking between any two activities only occurs at one level of density (usually the highest level of density). Allowing links to remain between activities at more than one level of density will cause inconsistencies in the calculation of scheduled dates, the critical path and float.

4.24.2   As the schedule is expanded from Low Density to Medium Density, the duration and timing of the original Low Density activity should be derived from the new underlying Medium Density activities (in most software tools this can be accommodated through the activity identifier coding structure – see Section 4.15). Therefore, the logical dependencies between the activities should also be at the Medium Density level (including links to later activities still at the Low Density level), and removed from the Low Density activities that have been expanded to Medium Density.

4.24.3   Similarly, as the schedule is expanded from Medium Density to High Density, the duration and timing of the original Medium Density activity should be derived from the new underlying High Density activities (in most software tools this can be accommodated through the activity identifier coding structure – see Section 4.15). Therefore, the logical dependencies between the activities should also be at the High Density level (including links to later activities still at the Medium Density or Low Density level) and removed from the Medium Density activities that have been expanded to High Density.

4.24.4   Unless there is a good reason to the contrary (in which case the reasons must be fully set out in the planning method statement), summary activities should not be linked. Where an activity has been expanded to a higher level of detail and now derives its duration and timing by summarising (or 'rolling-up') information from the more detailed activities, the schedule logic should be maintained at the highest level of detail. Therefore:

■   When a Low Density activity is expanded to Medium Density, the schedule logic should be created at the Medium Density level and removed from the Low Density activity (which now summarises the underlying Medium Density activities).

■   When a Medium Density activity is expanded to High Density, the schedule logic should be created at the High Density level and removed from the Medium Density activity (which now summarises the underlying Low Density activities).

4.25   Activity logic

4.25.1   The following are the logical possibilities for the identification of activity predecessors and successors.

Start-to-start

4.25.1.1   In the relationship shown in Figure 20, Activity B can start at the same time as Activity A but not before it.

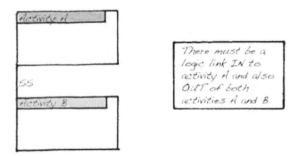

Figure 20 Start-to-start relationship.

Finish-to-finish

4.25.1.2   In the example of a finish-to-finish relationship in Figure 21, Activity B cannot finish until Activity A has finished. It implies that B can finish at the same time as A, but not before it.

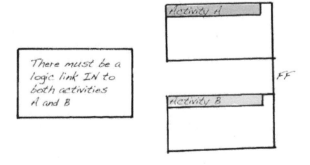

Figure 21 Finish-to-finish relationship.

Finish-to-start

4.25.1.3   The convention in Figure 22 shows the normal sequential relationship of one activity following another. For example, Activity B cannot start until Activity A has finished.

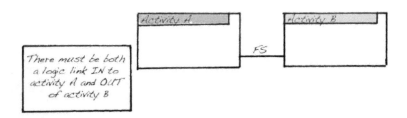

Figure 22 Finish-to-start relationship.

Start-to-finish

4.25.1.4   The convention in Figure 23 shows the unusual sequential relationship of one activity unable to finish until after the preceding activity has started. For example, Activity B cannot finish until after Activity A has started. It will be a rare occasion upon which this represents a reasonable logical progression, but it can be useful in managing the changeover of systems, for example, the use of 'temporary power' cannot end until the start of the 'permanent power supply'.

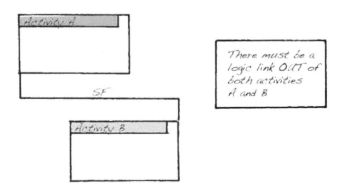

Figure 23 Start-to-finish relationship.

Computational inconsistencies

4.25.1.5   Wherever practical, the use of finish-to-start links is preferred. Using combinations of start-to-start and finish-to-finish links can produce unintended consequences when the schedule is changed. For example, in some situations, reducing the duration of a critical activity can cause the overall schedule duration to increase; in others changing the duration of a critical activity has no effect on the overall schedule duration. Schedulers need to be aware of these issues and how they are managed in the software algorithms and ensure the logic used is appropriate (see Section 4.30.7.33).

4.26   Lags

4.26.1   Time lags are generally used in scheduling to indicate a duration following the start or finish of a predecessor at which a successor may logically start, or a duration following the start or finish of a predecessor at which a successor may logically finish.

4.26.2   A lag is not an activity, nor is it a substitute for an activity. Scheduling software products do not always permit the user to attribute a calendar to a lag, but automatically attribute the calendar of either the predecessor or successor activity. In reality, the time lag represents an assessment of the time necessary to accomplish the necessary predecessor section of work, the quantity and identity of which may not necessarily be specified in the schedule. If it is not specified in the schedule, then the logical premise on which the lag is calculated must be identified in the planning method statement. As a general rule, the duration of lags should not exceed 50% of the shortest duration of the activities to which they are linked.

4.26.3   The various possibilities of the inclusion of lags are as follows.

Lagged finish-to-finish

4.26.3.1   In Figure 24, 'd' indicates a finish-to-finish relationship but with an imposed delay, that is, Activity B cannot finish until 'd' days (or whatever time period has been used) after Activity A has finished. This convention provides one of the means of overlapping the timing of activities. Instead of by reference to a time lapse, some scheduling-software products describe this as a relationship whereby a proportion of Activity B (either a percentage or a value) cannot finish until after Activity A has been completed.

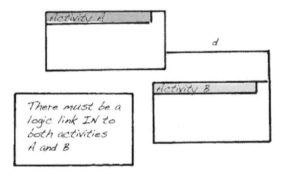

Figure 24 Lagged finish-to-finish relationship.

Lagged finish-to-start

4.26.3.2   In Figure 25, 'd' indicates a finish-to-start relationship between Activities A and B in which B cannot start until 'd' days after Activity A has finished. If the delay involves any form of activity by any party, the use of a lag is not appropriate and the activity must be described overtly in the schedule.

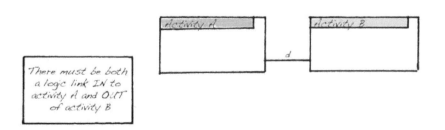

Figure 25 Lagged finish-to-start relationship.

Lagged start-to-start

4.26.3.3 Figure 26 indicates a start-to-start relationship with an imposed delay, that is, Activity B cannot start until the period 'd' has elapsed after Activity A has started. This convention provides one of the means by which the execution of activities can be overlapped. Instead of by reference to a time lapse, some scheduling software products describe this as a relationship whereby Activity B can start after a proportion of Activity A has started.

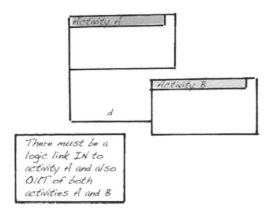

Figure 26 Lagged start-to-start relationship.

Lagged start-to-start and finish-to-finish

4.26.3.4 There may be occasions where a lag is required on both the start and the finish of related activities. This is achieved by the convention shown in Figure 27.

4.26.3.5 In this arrangement, Activity B cannot start until 'd' days after Activity A has started, and Activity B cannot finish until 't' days after Activity A has finished. For example, in the case of a pipeline, the activity 'lay pipes' cannot start until 'x' days after the start, or until 'y' days after the finish of activity 'excavate'. Thus, 'lay pipes' has a lag start of 'x' days and a lag finish of 'y' days after the start and finish date of the activity 'excavate'. Some scheduling software products describe this as a relationship whereby work cannot start or finish until after a proportion of the predecessor or successor has started or finished, instead of by reference to a time lapse.

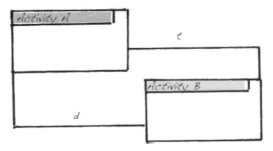

Figure 27 Lagged start-to-start coupled with lagged finish-to-finish.

4.26.3.6 A chain of three or more activities, with start-to-start and finish-to-finish driving lags, is called a ladder. Because of the propensity of many scheduling software products to give priority to driving finish-related logic over durations, if the duration of an activity in the ladder should change, these configurations can produce ridiculous results. Accordingly, ladders should be avoided unless the scheduling software being used is capable of producing a sensible resolution of the effect of logical inconsistencies.[22]

4.26.3.7   Some scheduling software products attribute the calendar of the predecessor activity to the lag, some attribute the calendar of the successor, and others permit an entirely different calendar to be allocated to the lag from that attributed to the predecessor and/or successor activities. Accordingly, in setting out the logical premises of the lag, its calendar should be identified.

Negative lag

4.26.3.8   Negative lags are also referred to as 'leads'. Negative lag is the relationship between a predecessor and its successor activity in which the successor is required to start a period of time before the predecessor. For example, a finish-to-start lag of 3 time units requires the successor to start three time units prior to the predecessor finishing. It is a logical construct that is impossible to perform; in the example, you do not know when the predecessor will actually finish until completion is achieved; the situation is more complex on other link types. Although not prevented by most scheduling software, negative lags should not be used; they can distort logic and falsify criticality, particularly as the schedule is updated.

Lags at Low Density

4.26.3.9   The use of lags is a useful scheduling technique at Low Density. Low Density schedules may legitimately contain a high number of lagged sequences, simply because at this density it is assumed that many of the activities will represent an entire section of work (e.g. 'construct building A'), and the flow to the next section (e.g. 'construct building B') requires only some of the work to be complete to allow the next activity to start.

Lags at Medium Density

4.26.3.10   For Medium Density schedules, durations will be shorter and the sequences will contain a higher number of finish-to-start sequences, but may legitimately contain some lagged sequences particularly with activities of longer duration.

4.26.3.11   At Medium Density, lags should be avoided wherever possible, particularly in relation to activities like concrete curing. The reason is that, although visible to the person who views the plan by using the scheduling software, it may not immediately obvious to those who view the schedule in hard copy or PDF.

4.26.3.12   Lags can be a useful scheduling technique at Medium Density levels in that they tend to reduce the number of activities it is necessary to illustrate. However, at Medium Density there should be less of a need to maintain lags previously inserted at low level, or to introduce new lags.

Lags at High Density

4.26.3.13   Without an accurate model of the planned intent, it will not be possible to forecast when activities are due to start or finish, or to identify, from time to time, whether any activity is critical to completion. It is thus of paramount importance that the logic applied to High Density schedules accurately illustrates what is required, and in High Density schedules lagged sequences will rarely be acceptable.

---

[22] One way of achieving this is to set the float calculation to 'most critical' where the software permits it.

4.26.3.14   Unless there is a good reason to the contrary (in which case the reasons must be fully set out in the planning method statement), lags should not be used at High Density levels. At this level, all activities should be identified sufficiently clearly for the resources following the schedule to understand at which point in the process each activity must be started and finished in relation to its predecessor.

4.27   Logical constraints

4.27.1   In most scheduling software products there are options available for manipulating the effect of logic with date or float-related constraints. The definitions of constraints and the way they are treated will differ between scheduling software packages. However, in all cases the effect of a constraint is to override what would otherwise be the logic of the network to which it is applied.

4.27.2   The effect of applying a constraint to an activity is to inhibit the activity from obeying any logic inconsistent with that constraint. The consequence of that may be that the schedule will be unable to predict, according to the logic, the dates upon which activities are due to start and finish, and it may also give a false impression of criticality. Accordingly, optional constraints, if used at all, must be used with great care and be justified in the planning method statement.

4.27.3   Legitimate uses of constraints include situations where the date for the commencement of an activity, or chain of activities may need to be established to model when resources are expected to become available at a specified point in time (referred to as a start-on or start-after constraint), or where events are to be scheduled at their latest possible dates (referred to as an as-late-as-possible constraint). Constraints can usefully be categorised as:

- flexible, in which activity start and finish dates will change according to any changes in logic and the associated resources;

- moderate, in which the activity start and finish dates will respect some changes in logic and the associated resources, but not all;

- inflexible, in which the activity start and finish dates are dictated solely by the constraint and will not change to reflect changes in logic and the associated resources.

Flexible constraints

4.27.3.1   *As-soon-as-possible*: This schedules the earliest possible start and finish dates for the activity. The late start and late finish are set equal to the early start and early finish. Float is eliminated from the activity and all its predecessors. In some software products this term is misapplied to describe the normal time analysis function of calculating early and late dates for each activity.

4.27.3.2   *As-late-as-possible*: This schedules the latest possible start and finish dates for the activity. The early start and early finish are set equal to the late start and late finish and float is eliminated from the activity and all its successors.

Moderate constraints

4.27.3.3   These should generally be avoided unless essential. They include the following.

4.27.3.4   *Zero-free-float*: This constraint will schedule the activity so that it finishes immediately prior to the commencement of the successor activity (often coupled with a defined lag representing a mobilisation period). This is an acceptable way of identifying, for example, the planned date of information release preceding a dependent activity, or for modelling 'just in time' materials deliveries.

4.27.3.5 *Finish-no-earlier-than*: This indicates the earliest possible date on which the activity can be completed, and the activity cannot finish at any time before the specified date. The occasions on which this is an appropriate constraint to apply will be rare.

4.27.3.6 *Finish-no-later-than*: This indicates the latest possible date on which the activity can be completed, but the activity can be finished on or before the specified date. The occasions on which this is an appropriate constraint to apply will be rare, but can usefully be used to identify contract completion dates and 'handovers', provided that the constraint is applied to a milestone that is not a part of the network. (see also Section 4.26.3.9 *et seq.*).

4.27.3.7 *Start-no-earlier-than*: This indicates the earliest possible date on which the activity can begin, and the activity cannot start at any time before the specified date. This is often used for the commencement of different phases of operation within a schedule where the phases otherwise have no logical dependencies, or access to areas is constrained by the terms of a contract.

4.27.3.8 *Start-no-later-than*: This indicates the latest possible date on which the activity can begin, but the activity can start on or before the specified date. The occasions on which this is an appropriate constraint to apply will be rare.

Inflexible constraints

4.27.3.9 Inflexible constraints may be usefully applied to a milestone to show contractual dates (for such things as employer-provided items, required interim handovers, contract completion, or approvals), provided that the milestone to which the constraint is applied stands alone as a contractual date mark and is not part of the network logic. In these circumstances, 'must-start-on' or 'must finish-on' constraints are reasonable constraints to use for the contractual dates. Where the milestone representing the contractual requirement is part of the network, no inflexible constraint should be applied. This is important as the milestone must be permitted to move dynamically with the logic of the network and demonstrate the logically achievable date for comparison with the contractually required date (see Figure 28).

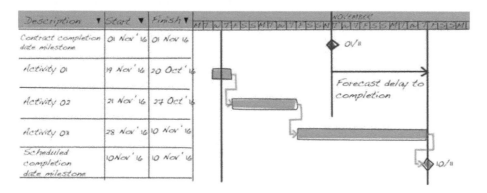

Figure 28 Contract milestones and network milestones.

4.27.3.10 In simple terms, any inflexible constraint used must not interfere with construction logic of the network. Inflexible constraints are thus not acceptable in a network, except in the schedule development stages where scenario-making related to assessing realistic duration and resource allocations to discrete sections of the schedule can be made with the use of inflexible constraints on start and finish of sequences. Such inflexible constraints must be removed from the operative schedule before it is put into use.

4.27.3.11   The inflexible constraints are:

- *Mandatory-start*: This constraint indicates the date on which the activity must begin. This takes priority over other scheduling parameters such as activity dependencies, lead or lag time, and resource levelling.

- *Zero-total-float*: The effect of this is to set the early and late dates to be the same and to render critical the activity to which it is applied, together with its controlling predecessors and successors.

- *Expected-finish*: This constraint identifies a finish date in the future, that is, to the right of the data date. Its effect is to change the duration of the activity to which it is applied to span between its early and expected finish date.

- *Mandatory-finish*: This constraint indicates the date on which the activity must finish. This takes priority over other scheduling parameters such as activity dependencies, lead or lag time, and resource levelling. Some scheduling software products permit finishing constraints to be assigned not only to activities but also to a section or the project as a whole. The effect of such a constraint is to determine the latest permissible finish date. This is usually set up in a way that is undetectable except by electronically interrogating the schedule setup.

Inflexible combinations of constraints

4.27.3.12   These also result in inhibiting the schedule from reacting to change and hence must not be used. They are:

- start-no-earlier-than in combination with start-no-later-than on the same date; this has the same effect as zero-total-float.

- finish-no-earlier-than in combination with finish-no-later-than on the same date; this has the same effect as zero-total-float.

4.28   Float

4.28.1   Float occurs in a critical path network as a result of the calculations made by reference to activity durations and logical sequence. Designated non-working periods (such as those identified as religious, industrial or statutory holidays, or weekends) are not float and should not be treated as such.

4.28.2   Consistent with the type and quantity of float, activities that have float will be able to absorb a degree of flexibility in their start and finish dates without affecting the critical path. However, because the available float does not remain constant (it changes after each recalculation of the network) and it is not normally preserved by contract for the use of any one party, the availability of float on some, or even on all, activities in a network should not be seen as a substitute for contingency planning.

4.28.3   The degree of float, which would otherwise exist, will be reduced as a result of resource levelling or the introduction of contingencies.

4.28.4   There are potentially many types of float that may be calculated in critical path networks. Their relationship to activities and to completion are illustrated in Figure 29. But in modern practice only 'total float' and 'free float early' (usually referred to as free float) are routinely considered.

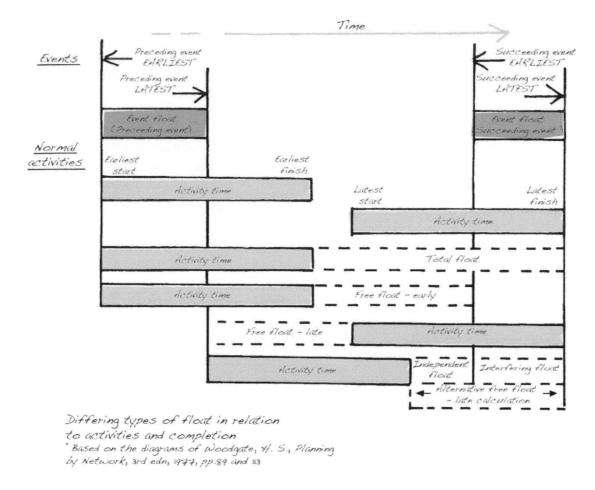

Differing types of float in relation
to activities and completion
* Based on the diagrams of Woodgate, H. S., Planning
by Network, 3rd edn, 1977, pp.89 and 113

Figure 29 Differing types of float in relation to activities and completion.

Free float

4.28.4.1   Free float is the period by which an activity can be delayed without affecting the start of any other activity. Free float is either a positive value or zero.

Total float

4.28.4.2   Total float is the period by which an activity can be delayed without affecting the relevant completion date or a constraint. Total float may be positive (the completion of the activity may be delayed, without delaying completion of the works), zero, or, if a constraint exists, negative.

Negative float

4.28.4.3   Negative float is an indication that an activity is logically scheduled to occur at a date later than an imposed inflexible constraint will permit.

4.29   Critical path

4.29.1   The critical path is the longest sequence of activities from commencement to completion of a key date, section, or completion of the works as a whole. It is that sequence of activities that will take the longest time to complete or, put another way, the sequence of activities that will determine the earliest possible finish in relation to a key date, sectional completion date or the completion date.[23] Hence, it is timely commencement and completion of those activities on that path which will secure completion of the key date, section, or the works as a whole on time. The critical path may be determined based purely on the durations and logic contained in a schedule, or may be influenced by the availability of resources and their pattern of use.

4.29.2   As activity start and/or finish dates change, or intervening events affect the sequence of work, the critical path has the potential to change, to add some activities that were previously not critical, and to remove from the critical path other activities which were critical. At each update or change to the schedule, the identification of the critical path and understanding what changes, which calculations and/or settings have affected, or caused changes to, the critical path, are very important first steps in the time management process.

4.29.3   In respect of each path, the scheduling software calculates the scheduled dates of activities by taking into account the durations of all activities and the logic between them. This is a three-stage process:

■   the first (or forward) pass starts from the beginning of the schedule and contin-ues to the end, calculating the earliest start and finish dates of each activity based on the duration and the logic of its predecessors;

■   the second (or backward) pass starts from the end of the schedule and regresses to the start, calculating the latest start and finish dates for each activity based on its successors; and

■   the final calculation is the float values for each activity, which are based on the start and finish dates.

4.29.4   When the scheduling software determines that the earliest and latest dates are the same for any activity start or finish, it indicates them and the logic between them as being critical and identifies that the start, finish, or the whole of the activity must be completed on time, in order to complete the work in the short-est time.

4.29.5   When the latest start and finish dates of an activity are later than the earliest start and finish dates, the scheduling software calculates the difference and identifies it as total float. This float can then be used to level resources or absorbed slippage during the course of the works, until the earliest and latest dates are the same, at which time the start and/or finish of the activity will become critical.

4.29.6   When the early finish date of an activity is earlier than the earliest start date of any of its successors, the finish of activity can be delayed for a period of time with-out affecting the start of any of those successors. This period is called free float. Very few activities have free float.

4.29.7   In the event that a completion date is fixed by a mandatory constraint (or combination of otherwise flexible constraints, which have the same effect) and the earliest start or finish dates of an activity are later than those fixed by the constraint, the scheduling software shows those activities to have less than zero total float (negative total float), indicating that delay to completion is predicted. If the project is to be completed on time, management actions are required to change the future work of the project to remove the negative float; these changes must be reasonable and achievable.

---

[23] The definition of Critical Path used in ISO-21500, Guide to Project Management, is: 'sequence of activities that determine the earliest possible completion date for the project or phase'.

4.29.8 Because different scheduling software products have different scheduling algorithms and, amongst other things, deal in different ways with conflicts, constraints and calendars, different software has the potential to calculate different float values and predict a different critical path from the same data. Accordingly, identifying the critical path requires some understanding of the methods and algorithms used by the software in producing the results of its calculations.

4.29.9 Defining criticality by reference to float values alone is not recommended. Total float values on a critical path may not always be the same because of the effects of resources, constraints, calendars and the scheduling method adopted. Additionally, float is diminished by the effects of resource levelling, which is also likely to affect the critical path calculations. The definition of the critical path recommended by the Guide is: 'The critical path is the longest sequence of activities from commencement to completion of a key date, section, or completion of the works as a whole'.

4.30 Schedule quality assurance

4.30.1 Because competent contract administration, timely completion and financial success of the project ultimately depend upon the integrity of the schedule, it is essential that the planning strategy and schedule are reasonable, rational and achievable. Achieving a reasonable, rational and achievable schedule is the responsibility of the project management team supported by the scheduler; however, validating these objectives have been achieved requires a review or audit.

4.30.2 Quality assurance audits are best performed by an independent party, not associated with the project or any of its participants; the absence of any implied knowledge helps to ensure that the right questions are asked and appropriate and understandable answers are given.

4.30.3 The first validation should be carried out at inception.

4.30.4 The assurance of quality and integrity in the schedule is achieved by both an initial validation of the development schedule followed by regular checks of schedule revisions and updates. In addition to reviewing the planning method statement and schedule, each audit should also check the contemporaneous records database for accuracy and completeness against work content and actual performance.

4.30.5 Subsequent and more detailed audits should then be carried out on the working schedule prior to any work stage commencing.

4.30.6 It should be borne in mind that the longer the period between audits, the longer the examination will take and the more serious are likely to be the consequences of any errors found. Accordingly, detailed audits should be carried out against revisions and updates at intervals no greater than two to three times the reporting period, depending upon the nature of the work being carried out.

4.30.7 Typically, the scope of validation will include examination of:

■ buildability

■ schedule content

■ schedule integrity

Review for buildability

4.30.7.1   This review typically encompasses an assessment of the appropriateness and the degree to which the schedule has addressed the planning method statement and, in particular, the following:

■   employer's requirements;

■   design and integration of specialist design;

■   quality specifications;

■   procurement;

■   construction;

■   health and safety; and

■   environmental matters.

4.30.7.2   The purpose of the buildability review is to minimise the possibility of intervening events occurring as a result of errors or omissions, and to assess the reasonableness of contingencies for the likelihood of delay occurring as a result of errors and omissions not being identified before work on site commences.

4.30.7.3   Experience in the discipline and project type to be assessed is vital to the success of this activity, since it is as a result of experience that potential areas of difficulty can be identified prior to construction and the causes of failure of previous projects of like kind can be avoided. Accordingly, a successful buildability review will involve consideration by a wide variety of construction professionals including, in relation to any defined area for consideration, those whose speciality is:

■   design

■   specification

■   production information

■   procurement

■   cost estimating

■   scheduling

■   information management

■   quality assurance

■   health and safety

■   environmental matters

■   sustainability

■   energy

4.30.7.4   Where the working schedule is conceived as a 4D model in a BIM environment, the virtual build can be used to assist in the buildability review, helping to identify errors and omissions, particularly missing design or scheduling information.

Review for schedule content

4.30.7.5   The schedule must illustrate a realistic and practical project plan showing how the work of the project is intended to be accomplished, in a form that is sufficiently accurate for its identified density.

4.30.7.6 It will include a review of the specific activities, the reasonableness of durations for the activities, and the planned sequence of work for the project. The logic or sequence of work activities should represent how the project is intended to be built and how the various activities are interrelated. The primary objective of a schedule content review is to determine that the project schedule contains the information necessary to render it fit for its intended purpose and is accurate, logical and achievable.

4.30.7.7 While automated checking tools are readily available to assess the technical correctness of the schedule, these tools cannot validate the reasonableness or 'sense' of the schedule. The purpose of this review is thus to establish that, for any given density, both the planning method statement and schedule adequately describe what is intended as a dynamic time model.

4.30.7.8 The review will include examination of:

- planning strategy, means, methods and assumptions;

- work breakdown structure;

- coding structure;

- estimates and calculations of resources, productivity and durations;

- engineering logic, resource logic and preferential logic;

- contractor, subcontractor, work package logic, production and space restraints;

- constraints, milestones and performance duration;

- cost codes, budgets and earnings;

- calendars;

- design fabrication, procurement and delivery lead times;

- submission and approvals schedule;

- risk register and contingencies; and

- the critical path to each key date, section, or completion date of the works as a whole.

4.30.7.9 Where as built data is incorporated into an update, the review will also include validation of the schedule against the progress records:

- submission and approvals register;

- as-built and progress data;

- performance and productivity data; and

- cost budgets and earnings.

4.30.7.10 Where intervening events have occurred, the review will also include validation of:

- event register;

- event synopses and fragnets;

- impact methodology and calculated effect;

- the critical path to each key date, section, or completion date of the works as a whole;

■ recovery and acceleration strategy and proposals; and

■ recovery and acceleration monitoring.

4.30.7.11   Where the working schedule is conceived as a 4D model in a BIM environment, the 4D model can be used to assist with the review of schedule content by ensuring that adequate information is included in the activity descriptions and that all design items are considered in the schedule.

Review for schedule integrity

4.30.7.12   The purpose of this review is to establish that the schedule and planning method statement are completed to a density appropriate for their use, and that the schedule will respond dynamically to change.

4.30.7.13   Many scheduling software products contain facilities for distorting the schedule or for concealing its inadequacy. In inexperienced hands this can result in schedules that conceal deficiencies in logic, durations, progress or content, which may not be uncovered until it is too late to make the necessary corrections.

4.30.7.14   Accordingly, the purpose of a review of schedule integrity is to ensure that the schedule used as the time model can function effectively as such and can be used reliably and safely, at any time, to predict the consequences of actions and inactions.

4.30.7.15   The integrity of the schedule is of paramount importance in predicting consequences because it is the skeleton for the calculation of the effect of any intervening events added to it, and the baseline from which efficacy of corrective action can be estimated. If the working schedule does not react dynamically to change, or does not react logically, the calculation it produces will be of little help in identifying cause and effect, or predicting the future conduct of the work.

4.30.7.16   Accordingly, the process of reviewing schedule integrity is to interrogate the schedule to identify any faults that might inhibit its use as a time model. In principle the review will include investigation of:

■ a level of detail appropriate to the level of density;

■ density logic;

■ constraints and constraint types;

■ activity logic, including:

  ○   open ends

  ○   long lags

  ○   negative lags

  ○   ladders

  ○   scheduling options

  ○   critical paths.

4.30.7.17   Where the working schedule is conceived as a 4D model in a BIM environment, the 4D model can be used to assist with the review of schedule integrity by showing the virtual building process. This can help to establish whether the schedule can function physically and can be used reliably and safely to predict the consequences of actions and inactions, and that the logic makes sense.

Review for constraints

4.30.7.18   Any chain of activities that is identified as critical but does not start at the beginning and finish at the end of a schedule is likely to be illustrated as critical only because of a manually applied constraint which has dictated that effect.[24]

4.30.7.19   Because such constraints distort float calculation, and hence criticality, all inflexible constraints and combinations of constraints which produce inflexibility should be removed.

4.30.7.20   Other manually applied constraints should be removed and replaced with logic links wherever possible.

4.30.7.21   Constraints may be applied to a milestone to show contractual dates (for such things as employer-provided items, required interim handovers, contract completion, or approvals), provided that the milestone to which the constraint is applied stands alone as a contractual date mark and is not part of the network logic.

Review for open ends

4.30.7.22   Open ends are sometimes referred to as dangles. Activities that have either no predecessors to their start (leading open ends), or no successors from their finish (trailing open ends), must be identified and corrected.

4.30.7.23   Because many scheduling software products identify as open-ended only those activities with no predecessors or successors, rather than those activities with no predecessor to the start, or no successor to the finish of the activity, either a careful manual review or the use of automated checking tools is recommended.

4.30.7.24   Where the activity concerned is the starting milestone identifying commencement of the project, there is no logical predecessor; other leading open ends should be identified and logic added to remove them where possible.

4.30.7.25   Ultimately, completion of every activity is a precondition for final completion and all trailing open ends must be removed.

Review for long lags

4.30.7.26   Lags represent an impediment to the progress of a successor. The lag should represent a portion of a predecessor, which impedes the successor's progress and not an unscheduled activity (such as concrete curing).

4.30.7.27   Lags should not be used to represent work by others such as off-site procurement, or for activities by parties other than the schedule's author, or to represent unknown scopes of work to be detailed as the information comes to hand. These are practices to be avoided and the logic corrected with the inclusion of appropriate activities.

4.30.7.28   Lags cannot be statused during schedule updates; accordingly a lag should be replaced with an activity wherever possible.

4.30.7.29   If any lag is longer than half the duration of the shortest activity to which it is linked, it is usually an indication of faulty logic and should be thoroughly investigated.

---

[24] See 'Inflexible constraints' at Section 4.27.3.9 et seq.

Review for negative lags

4.30.7.30   These are used to create overlaps, where the timing of the successor is dependent on the future completion of the predecessor: a logical impossibility.

4.30.7.31   Negative lags are fatal to a time model, which depends upon its logic for the result it displays; they distort float calculations and hence criticality.

4.30.7.32   Negative lags should be removed and replaced with suitable logic. Typically, the predecessor activity should be split into two activities to resolve the logic into a finish-start, coupled with a start-start and finish-finish.

Review for ladders

4.30.7.33   A ladder is a series of three or more activities, all of which are linked start-to-start and finish-to-finish with driving or near-driving relationships. Because this overlapping type of structure is so common and useful in construction projects, particularly in schedules of Low Density and Medium Density, understanding how the software deals with this type of situation can be critical to the validity of the time model.

4.30.7.34   Depending upon the software used and the way it is configured, the shortening or lengthening of the duration of an activity in a ladder may produce ridiculous results, catapulting its successors forward in time (see Figure 30).

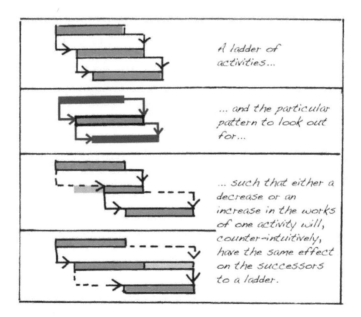

Figure 30 The problem with ladders.

4.30.7.35   The reason for this is inconsistency between driving, start logic, activity duration and driving, finish logic; many scheduling software products dictate priority to driving, finish logic.[25] On the other hand, some software products provide the alternative facility for taking activity durations as a priority by dictating them to be intermittent instead of contiguous, or have the facility for designated 'ladder activities' such that they act in a feed-type manner so that progress in successor ladder activities is proportional to predecessor activities.

---

[25] This can be avoided by setting float calculation to most critical, if the software provides for it.

4.30.7.36   Ladders should be investigated and, where the software does not handle them adequately, they should be broken into their constituent activities in more detail.

Review for scheduling options

4.30.7.37   The scheduling options adopted within the scheduling software must be reviewed in order to gain an understanding of the way the schedule operates and makes its calculations.

4.30.7.38   Identify whether the activity durations are based upon interruptible or contiguous activity durations. The normal default method of calculating activity duration is contiguous.

4.30.7.39   Check whether the scheduling method is set to progress override or retained logic.

4.30.7.40   Consider whether lag is calculated from early start or actual start.

4.30.7.41   Review the method of calculating total float.

Review for critical paths

4.30.7.42   A review of all critical paths from commencement to completion of each key date, section, phase and the project as a whole is essential.

4.30.7.43   If an unbroken path cannot be traced from completion back to commencement, it will be because there are constraints in the schedule distorting the logic, or incomplete logic.

4.30.7.44   Missing logic should be added and, if there are constraints in the schedule, these should be removed and replaced with logic, if at all practicable to do so.

4.30.7.45   The percentage of activities on any critical path will change according to the density of the schedule. Commonly, at Low Density 50% of the schedule could be critical, whereas in Medium Density around 15% of activities being critical to completion will be more usual. If there is significantly more or less than these heuristics, the reason should be investigated.

4.30.7.46   Ideally, all the critical paths should be reviewed and discussed in detail with the employer, design team and project management team to gain understanding, both before and after any changes to the schedule have been undertaken.

5 Managing the dynamic time model

5.1 Introduction

5.1.1 In summary, management of the DTM is the process of maintaining the schedule during the delivery of the project which, in turn, involves:

■ review and revision of the assumptions used to produce the schedule;

■ collection of progress records;

■ monitoring the work in progress;

■ updating the schedule;

■ identifying intervening events;

■ impacting them on the schedule;

■ implementing recovery and/or acceleration;

■ revising the schedule as better information becomes available; and

■ revising the planning method statement to record what has changed and the reasons for it.

5.1.2 The employer and its professional team are entitled to know what the contractor has achieved to date, and whether it is on target to complete by the various key dates, sectional completion dates and the contract completion date. The working schedule is thus not just the schedule of the contractor's intentions, but a fundamental management tool for the employer and the design team. It provides:

■ the essential information regarding the timing and interface between their continuing obligations and the works; and

■ the calculating mechanism in the event that the contractor becomes entitled to prolongation costs and/or an extension of time.

5.1.3 For its success, management of the DTM depends upon the quality of the working schedule and its ability to absorb and integrate the information collected during the management process. Hence, the party operating the scheduling software and maintaining the schedule has a duty to ensure that the schedule is robust and as accurate as it can be.

5.1.4 As better information becomes available, the schedule must be revised to incorporate it. Revision is not updating, nor is it the impacting of intervening events. It is the process of adopting a change in the assumptions and/or calculations previously used in the planning of the contract works, as better and more precise information becomes available during the course of the project.

Guide to Good Practice in the Management of Time in Major Projects: Dynamic Time Modelling, Second Edition. The Chartered Institute of Building.

5.1.5   The relationship between the different levels of schedule revision, monitoring, updating and impacting in the schedule management process is illustrated in the flow chart in Figure 31.[1]

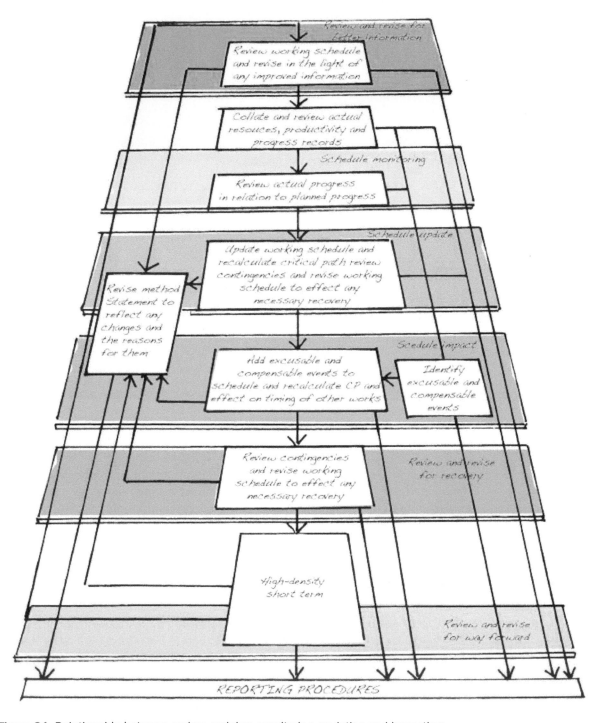

Figure 31 Relationship between review, revision, monitoring, updating and impacting.

5.1.6   The purpose of updating the schedule is to take into account progress actually achieved to ensure that the working schedule accurately takes account of the effect of what has gone before and, as a result, predicts what is to be done next,

---

[1] The evolution of the dynamic time model is also described in detail in Part 3 and in the flowchart in Appendix 4.

when it is to be done, and what resources are to be involved. This is essential for competent resource management.

5.1.7   The time management process must include accurate record-keeping, particularly those records regarding progress. Progress records provide essential data for understanding what has occurred to date and underwriting forecasts, calculating effects of change and lessons learnt.

5.1.8   Provided that the schedule is up to date, it can then be used for, amongst other things, identifying the occurrence and ultimate effect of an intervening event:

■   on progress, and the future conduct of the work;

■   on one or more key dates, sectional completion dates and the completion date; and

■   on one or more other contracts and subcontracts.

5.1.9   All major projects will be affected by an intervening event at some time. However, whilst it is normally not possible to predict when intervening events will occur, or what they will affect once they have occurred, time cannot be managed effectively unless the consequences are calculated promptly and appropriate action taken in good time.

5.1.10   When relevant, the DTM can also identify and quantify (in relation to the terms of the conditions of contract) the rights of the parties to extensions of time and liquidated damages, and time-related compensation for disruption and prolongation.

5.1.11   If the works are being, have been or are likely to be adversely affected by an intervening event, the DTM can also be used to manage the effects of the change, for example, by the allocation or reallocation of contingencies, identifying the effects of alternative processes which might be adopted by way of mitigation, recovery or acceleration, and as the model for any supplemental agreements as to change in process.

5.1.12   Finally, the as-built schedule can be used for identifying the actual sequence, resource and durations which can be used in the future to improve the predictability of the model.[2]

5.2   Data communication systems

5.2.1   Whilst a simple database can be used primarily for the purpose of recording resources expended in relation to the scheduled activities, it is possible to develop the simple model to provide a database suitable for maintaining records of all project documentation, notices, test records, correspondence, events, issues and the like (see Figure 35 below).

5.2.2   Where the database is network-based, it can be used by others and on many projects using a wide area network. A form like this can be operated by an extranet with special access rights to a defined user group, or on a company's internet site with defined user rights.

5.2.3   However, networking is not the only means of interfacing with the database. This form may be linked directly to a server storing records from many sources. If email forms are used to interface with the database, an electronic data receipt regime will be required to transfer the emailed data into the database. For this, typically, a dedicated email address is established with automated procedures that are triggered by the incoming mail. The database can have all the integrity restraints of the network

---

[2] See 'Feedback and benchmarking' at Section 6.6.

database and can automatically email replies to warn of potential errors and duplications, requesting revisions and confirmations.

5.2.4   Cloud-based systems (a form of networking including many of the attributes outlined above) are also becoming much more popular in project-based work. It gives an organisation the flexibility in choosing computing resources, hardware and/or software, to suit its particular needs. These reside on a remote machine and are delivered to the end user as a service over a network, currently the most common being the internet. The user entrusts the data to a remote service, on which the user has limited or no influence.

5.2.5   There are four main models of cloud computing:

■   Infrastructure as a Service (IaaS);

■   Platform as a Service (PaaS);

■   Software as a Service (SaaS);

■   Network as a Service (NaaS).

5.2.6   The main advantage in cloud computing, of whatever type, is accessibility, in that no matter the type of device or the geographical location, subject to user permissions and the like, data input and output is readily available to suit any particular user's needs.

5.2.7   Where the design is produced in a BIM environment, the use of a common data environment (CDE) for containing and managing the data has significant advantages over other methods of data control.

5.2.8   The CDE acts as a repository for and single source of access to, for example:

■   documentation

■   the graphical model

■   the dynamic time model

■   cost data

■   photographs

■   drawings

■   voice recordings

■   other non-graphical data

5.2.9   A CDE may be provided by a discrete project server, via an extranet, a file-based retrieval system, or by any other suitable toolset.[3]

5.3   Building information modelling

5.3.1   A 4D model can assist the management of time by representing, visually, in the 3D virtual model the degree of progress made at a point in time. It can do this relatively quickly and transparently. Accordingly, it encourages the engagement of all project team members, identifying responsibilities and assisting with the absorption and integration of information during the management process.

---

[3] PAS-1192:2: Specification for information management for the capital/delivery phase of construction projects using building information modelling.

5.3.2 BIM provides an environment for close coordination between those undertaking work and management and helps to understand the transit from Medium to High Density, leading to accurate reflection in the 3D virtual model of what occurred on site.

5.3.3 4D also acts as an electronic database for records, which can be linked to or embedded into the software. These records can be sorted and filtered, but care should be taken not to make the model too data-heavy. Therefore, external databases which support CDE and link to 3D/4D models can often be used to better advantage.

5.4 Record-keeping

5.4.1 Records that cannot be retrieved are useless. Accordingly, the process of record-keeping is inseparable from the process of record retrieval. It follows that in order to identify an adequate means of keeping records in the first place, consideration must be given as to how they can be retrieved and used.

5.4.2 The increasing speed and sophistication of databases and spreadsheets have provided the industry with the facilities for sorting and filtering data into specific reports at the press of a button, and it is now no longer acceptable to keep records on paper alone. It follows that unless records are kept as database records in the first place, they must be rekeyed as database records before they can be retrieved and used.

Spreadsheet-recorded data

5.4.3 The purpose of using a spreadsheet is to facilitate simple filtering and evaluation and reporting, with the opportunity for electronic import into a database if required.

5.4.4 If a spreadsheet is used, the records must be kept on a standard software product and be laid out so that the information can be automatically imported into a database either at initiation, or subsequently, without the need for rekeying.

5.4.5 This requires that:

- each activity identifier occupies a separate line;

- the information relating to that identifier occupies the same line as the activity to which it refers; and

- each item of information occupies a separate cell consistently from report to report.

Database-recorded data

5.4.6 Major projects require significant record management, which can only be properly achieved by keeping the records in a database to which all those needing to see or use the information can have access. It is the relationships within a database that facilitate the organisational and filtered view and retrieval of the information (by way of the fields and values within the fields of each record), which, when well designed, make databases the most functional of all record-keeping methods.

5.4.7 The advantages of keeping records in a well designed database are:

- records are stored in an accessible electronic format providing wide access for record entry and review;

■ every record stored is attached to appropriate identifiers, enabling organisation, filtering, targeted review and reporting;

■ access to the record identifier and content can be controlled to minimise erroneous record entry;

■ background checking can be established to prevent record errors such as duplicate entries and excess hours for labour or plant;

■ information is entered only once but can be grouped and reported to suit many output requirements; and

■ the records themselves can be readily monitored to ensure that they are being maintained in the manner required.

5.4.8   However, the overriding advantage of maintaining records in a database is that their content forms one unified source of information and, as a result, all reports, extracts and summaries produced by the database will be based upon the same data and thus have mutual integrity.

5.4.9   Figure 32 illustrates the relationship figure of a simple database for recording project activities. This figure illustrates a database facilitating a record of data kept in three fields: 'Activity', 'When it was done', and 'What resource'.

5.4.10   Two fields, 'Schedule Activities' and 'What was done', serve to provide the data for each of the fields called 'Activity ID', 'Activity description', 'When it was done' and 'What resource', which then contribute information to a table called 'What, when and resource'.

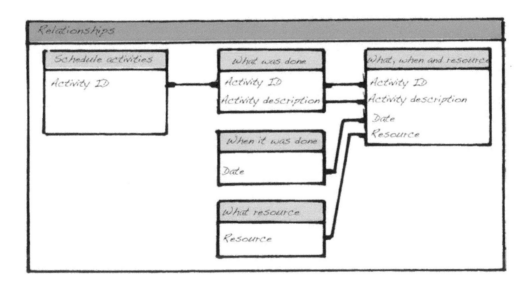

Figure 32 Simple database relationship.

5.4.11   Each of these tables is very similar to a spreadsheet. For example, when opened in the database, the table of 'What, when and resource' appears in the form illustrated in Figure 33.

5.4.12   The database also contains review and reporting features enabling searching and grouping of data to any number of declared criteria.

5.4.13   The database is designed to offer a user-friendly form for record entry. A simple form for making an entry is shown in Figure 34, while Figure 35 shows a simple database structure that might be populated by such forms.

What, when and resource				
Activity ID	BOOI70			
Activity description	Reduced level dig			
Date 17 June	Task	Resource	No. of	hours
	Cart to tip	Dumper	1	4
	Cart to tip	Machine operator	1	4
	Excavate	Excavator	1	4
	Excavate	Machine operator	1	4

Figure 33 What, when and resource table.

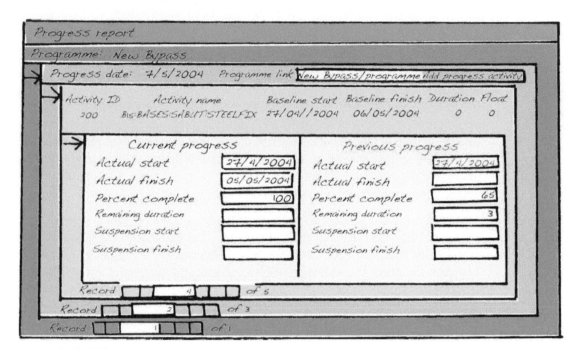

Figure 34 Simple data input form.

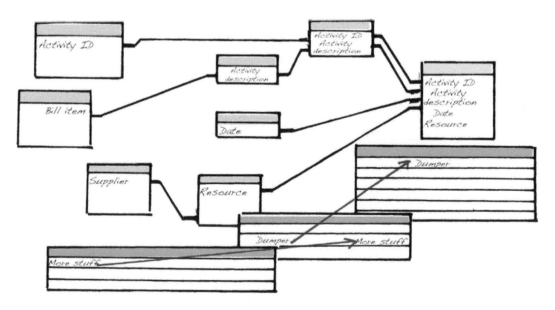

Figure 35 A database containing supplier data and bill of quantities data.

Record types

5.4.14   The records which generally have to be kept for successful time management fall into a number of categories:

- progress records

- quality control records

- information flow records

- change control records

- environmental records

5.4.15   Ideally, they should be kept in the same record control system and interrelated through a database; otherwise, unless the separate databases are electronically integrated, double entry will be necessary to some extent, and the likelihood of error in data input will increase.

5.5   Progress records

5.5.1   Records of progress actually achieved will be used for identifying the start and completion dates of activities, the degree of progress achieved from time to time, and for identifying the productivity actually achieved by particular resources. They can also be used for verifying productivity trends, the timing of the constituent parts of intervening events and identifying lost productivity as a result of disruption. In other words, progress records are the lifeblood of effective time management.

5.5.2   A consistent approach should be adopted for the gathering and collating of progress data, and at each progress update the data-gathering process should be carried out following the same sequence. Because the progress records are the source from which the schedule will be updated, the scheduler should have primary responsible for the collection of the data, so as to support an audit trail from the observation of a fact through to the inclusion of the information in the updated schedule.

5.5.3   The optimum frequency of data collection is not related to the schedule update period, or monitoring frequency, or the required reporting period, but will usually be related to the number of activities and the complexity of their interrelationships being carried out from time to time.

Progress record content

5.5.4   The records to be kept are those that will help in the management of the works going forward and the establishment of the facts surrounding the work which has gone before. In effect, these are records concerning:

- information flow

- progress

- photographs

- diary records

- note/log books

- minutes of meetings

- records of transactions (receipt of information, materials, labour, plant and goods, etc.)

- weather, industrial action and other issues
- third-party issues
- quality control
- change control
- productivity achieved

5.5.5   Unless the right information is kept in relation to the purpose for which it is to be used, no matter how accurate it is, or how well it is presented, it will be ineffective and of little use. It follows that the purpose for which the records are to be used will be the overriding factor as to their content. However, there are some types of data that are essential to any record, no matter for what purpose it is to be put. These are:

- activity identification data code
- activity description
- date of record
- resource used
- activity start and finish dates
- the author of the record
- progress data

Activity identification data

5.5.6   There must be a relationship between the planned activity (if any) and the records of work done to carry out the activity. This will ordinarily be the unique activity identifier code (activity ID) created when the activity was planned.

5.5.7   If the work being recorded cannot properly be allocated to a planned activity, it must be allocated to an intervening event and identified separately with a new event-related activity ID.

Activity description

5.5.8   Whenever possible, the description of the work being done should follow the description of the work in the High Density part of the schedule. If it does not follow it, it will be because either the activities planned are in insufficient detail to be properly recorded, or the work being carried out was never planned to be carried out and constitutes an intervening event.

Date of record

5.5.9   The date and, if relevant, the time at which the record is made must appear within the record. Unless the content of the work achieved can be related to the date it was recorded to have been achieved, the recorded proportion of work completed is meaningless. The date of a record also adds credibility to a recorded activity start or finish date.

The resource

5.5.10   The record must identify the quantum of resource being used to achieve the work. In so far as work in High Density is scheduled according to the resources to be applied, without the identified resource actually used, the record will be ambiguous and, for some purposes, may be useless.

Start and finish dates

5.5.11   The actual date an activity started and finished must be recorded.

Author of the record

5.5.12   Whilst for many purposes the author of a record may be self-evident, that will not always be so, and in the event of discrepancy, the identity of the party making the record can be very important.

Progress data

5.5.13   If an activity is started and completed in its entirety in a reporting period, the work done will be self-explanatory. However, if it is not, then the quantity of work achieved in the reporting period, together with the date upon which the record was made, must be identified.

5.5.14   Depending upon the facilities offered by the scheduling software, progress can usually be identified in four different ways:

■   remaining time;[4]

■   expired time;[5]

■   proportion of effort expended;[6] and

■   quantum of work done.[7]

5.5.15   Bearing in mind that it is always the High Density part of the schedule that is to be updated with progress achieved, and when that work was planned, the activity durations were calculated by reference to the resources intended to be used and their planned productivity, it is important to record the as-built data in each unit time period for:

■   the activity pursued;

■   the type of resources used;

■   the quantity of resources used; and

■   the quantity of work achieved.

5.5.16   Whenever a repetitive cycle is scheduled, great care must be taken to ascertain and check the resources planned to be applied, the gang strength, quality control, and anticipated productivity. If it is possible to run trial cycles by way of establishing a benchmark before construction starts, so much the better. However, in practice it is likely that, in many cases, benchmarking will have to take place over a period of time with different design teams or work gangs to identify differences between ultimate productivity and learning curves with resources of differing abilities.

5.5.17   In principle, apart from the basic data (see Section 5.5.13 above), the information which must be kept in each unit time period will produce the answers to the following questions, on an activity-by-activity basis:

■   What was done?

■   How much was done (the quantity of work completed)?

---

[4] By reference to the expected time needed to complete.

[5] By reference to the expired period in relation to the planned duration.

[6] By reference to the percentage of work done in relation to the whole.

[7] By reference to the measurement of the resources used and resources remaining in relation to the measurement of work performed).

- What duration is remaining to complete the activity (hours, days or weeks)?

- Who did it (the labour resource)?

- With what was it done (the plant and material resource)?

- When was it done (the date and timing of the activity)?

- Where was it done (the location in which it was carried out)?

- How was it done (the process adopted)?

Quality control records

5.5.18   A system of records that prompts the inspection, checking and testing of the works is often a requirement of the specification, and it may be required to be formally submitted as part of the employer's final acceptance procedures. Typically, such records will identify:

- subject matter;

- specification requirements;

- test date;

- testing supervisor;

- test results;

- deficiencies identified;

- corrective action taken;

- date upon which corrective action commenced and completed;

- sign-off by test supervisor and certificate of compliance.

5.5.19   As with other information, quality control documentation can be managed by a proprietary document management system (DMS) product. However, if these records can be integrated or synchronised with the database used for the progress data, it will be an advantage.

Information flow records

5.5.20   For the efficient exchange of information it is necessary to record in relation to any informational transaction:

- unique information-item identifier;

- its subject matter;

- the source of the information;

- what actions are needed in response;

- who will take this action; and

- when the action is commenced and completed.

5.5.21   There are many DMS products which can be used to accumulate and report upon information related to this type of data. However, if these records can be integrated or synchronised with the database used for the progress data, it will be an advantage.

5.6 Updating the schedule

5.6.1   Updating the schedule with progress data is not progress monitoring, nor is it schedule revision. It is simply the addition of as-built data to the relevant schedule and the recalculation of the critical path based on the addition of data from the progress recording system of the progress actually achieved by the data date. The records of progress are used to add as-built start and finish dates to those activities which have achieved either status, and progress data to those started but incomplete at the defined data date. Where work has progressed out of sequence, adjustments to logic will be necessary in order to ensure that the updated schedule accurately reflects the real state of progress, and work scheduled for the next day can realistically be expected to be worked on.

5.6.2   Updating a schedule is essential to the management of the dynamic time model; without it, the schedule is merely a target against which, historically, failure to make progress (for whatever reason) might be plotted. By updating with progress against the progress records and recalculating the critical path, the schedule becomes a dynamic model by which:

■   predictions can be made;

■   problem issues can be identified early;

■   mitigating recovery and acceleration can be planned and implemented; and

■   the future conduct of the works can be effectively managed.

5.6.3   The prompt identification of disruption and delay to progress is vital if the consequences are not to result in loss of quality, cost overruns and delayed completion. Once identified, difficulties can always be addressed, their potential consequences calculated, and strategies implemented to avoid or reduce those consequences.

5.6.4   An essential part of the updating process is recalculation of the critical path following the addition of progress data. The recalculation of the critical path identifies the start and finish of activities to be commenced, those which are then on the critical path, and provides a cogent baseline against which the effect of intervening events can be calculated.

5.6.5   When the schedule is properly updated to a data date with progress achieved:

■   resource planning is rendered more reliable because past and current experience of productivity can be better used to forecast future trends;

■   the impact of change can be accurately predicted against the model, whether it be to an activity duration, sequence (logic), or resource (money, labour, plant and materials);

■   The effect of changes to construction activities can be better calculated with 'what-if' scenarios, enabling the project team to choose the most efficient sequences;

■   potential problems are identified much earlier, with resulting increased time available to mitigate any risks, deal effectively with any issues from any 'crystallised' risks, and thus improve the likelihood of maintaining the project's time and resource targets; and

■   the cause and effect of disruption can more readily be identified and its consequences managed.

5.6.6   Many of the PC-based project planning software products permit the user to bypass inconsistencies between the network logic and the progress data in the schedule update process, according to the user's choice of progress overriding the inconsistent network logic or retaining the network logic and ignoring the effect of the inconsistent progress data.

5.6.7   However, both options produce a misleading result and neither is an acceptable alternative to properly updating the schedule with accurate logic of the progress actually achieved.

5.6.8   In summary, the high quality management information delivered by the updated working schedule encourages its use by the project team for management purposes.

5.6.9   For general purposes, there should be a direct correlation between the timing of the schedule update and the reporting cycle. In other words, the data date of the updated schedule should match the reporting requirements. Legitimately, however, there may be an increase in updating frequency depending upon the construction phase and/or criticality of the work in progress.

5.6.10   Prior to commencing the schedule update, the schedule should be backed up so that, prior to update, the schedule can be archived for comparative purposes and the updated schedule can then be maintained as the new operative schedule.

5.6.11   During the schedule update, an audit trail should be maintained of the source of the as-built data.

5.6.12   On completion of each update, the critical path should be recalculated and recorded and the resultant changes to the critical paths noted and recorded in the planning method statement.

5.7   Schedule review and revision

5.7.1   Review will not necessarily mean that a revision will follow, but experience tells us that unless a major project is in the last few weeks before completion, it is very likely that revisions to some part of the planned intent will need to be made as a result of a review.

5.7.2   An effective dynamic time model management process will generally involve review and revision based on:

■   improved information about what is contractually required;

■   resources and productivity as schedule density increases.

5.7.3   Reviewing and revising a schedule must be the result of a formalised management and governance procedure involving the project team.

5.7.4   Prior to commencing any review, the schedule should be backed up so that, prior to revision, the schedule can be archived for comparative purposes and the revised schedule can then be maintained as the operative schedule.

5.7.5   During revision, an audit trail should be maintained of all changes and the reasons for them, which should be recorded in an update of the planning method statement.

Review for better information

5.7.6 Typically, those areas of the schedule which require attention are those concerning:

- design
- procurement
- work content
- remaining short-term work

Better design information

5.7.7 Check, review and revise, where necessary, those activities concerning:

- design methodology
- specialist input
- drawing and information control
- submissions
- approvals

Better procurement information

5.7.8 Check, review and revise, where necessary, those activities concerning:

- procurement
- work package definition
- bill of quantities production
- tender appraisal
- contract and mobilisation

Refinements to work content

5.7.9 Check, review and revise, where necessary, those activities concerning:

- prime cost sums
- provisional sums
- approximate quantities
- plant availability
- construction methodology
- construction resources and productivity
- testing and commissioning

Review for short-term work

5.7.10 Check, review and revise, where necessary, those activities in the resourced short-term, High Density parts of the schedule.

5.7.11   Resource review (e.g. manpower, plant and materials, etc.) has an important part in ensuring that a project's objectives are met most efficiently and effectively. Resources should be reviewed in terms of:

- ■ suitability

- ■ type

- ■ availability

- ■ output

- ■ cost

5.7.12   Whilst policy decisions on the type of resources intended to be used are best made at the early design stages of a project, reviewing resources at any stage can also help in producing the most efficient and time-effective schedule and is essential for the development of the short-term, High Density part of the schedule.

5.7.13   In order to transition from Medium to High Density, close coordination is required between those who will actually carry out the work and the management team. It is essential that the work is accurately scheduled to reflect the workforce's intent and that the workforce intends to follow the work sequence with the resources allocated. Accordingly, before completing this review it is essential that the parties to carry out the work express their confidence that, in the absence of any intervening event, they can:

- ■ produce the resources required;

- ■ achieve the productivity envisaged; and

- ■ work to the planned sequence with the required interfaces.

5.7.14   When working with resources, it is important to understand how the software analyses resource utilisation. Of particular concern should be whether, in the event of overload, the completion date will remain unchanged and resources will be increased above the indicated limit, or whether the completion date will change to reflect the time required to complete the project with the allocated resources.

Change in methodology

5.7.15   The implications of making changes in methodology should be carefully explored and considered in terms of the likely effects on time, cost, quality, risk and conflict with the contract conditions (e.g. a change from a traditional design prepared under the direction of the employer with a build-only construction contract to one of design and build is likely to affect all phases of a project).

Repetitive activities

5.7.16   Trends in the measured productivity achieved should be used to verify the planned schedule for the remainder of those activities. If a discrepancy is found, changes should be made to the working schedule to correct the discrepancy in the planned work going forward. In relation to an activity such as piling, for example, each rig's productivity should be analysed (in order to ascertain the optimum pile cycle achievable and the effect of any prior departures from it). The demonstrable, achieved productivity cycle can then be used to verify the activity durations planned for the remaining piling works.[8]

---

[8] For examples of the benefits of this in practice, see Appendix 5.

5.7.17   If it should be found that the productivity achieved is insufficient to maintain the schedule, changes can then be made in good time to the planned resources and/ or to the sequence of the works in order to bring the work back on schedule.

5.7.18   One of the advantages of this repetitive sequence review is that in the case of an occurrence which disrupts productivity, the benchmark proven optimum productivity will be the best possible baseline against which to calculate the effects of the disruptive event.

Change in activity descriptions

5.7.19   The scope of activity descriptions will vary depending upon the density at which the schedule is prepared. At Low Density the schedule will have very broad overarching descriptions, whilst at Medium Density and High Density, schedules will have more detailed, discrete activity descriptions related to the activities in a single work package. Any review and change should, therefore, take these various densities into account, ensuring that the descriptions fit the purpose for which they are intended.

5.7.20   Activity descriptions should be reviewed with the following questions in mind:

■   Does it accurately describe the activity so that it is unambiguous in its scope and meaning, is easily understood and consistent with other descriptors?

■   Does the activity or group of related activities match the planning method statement?

■   What is the reason for any change to the activity description?

Change in estimated activity durations

5.7.21   Estimated activity durations must be regularly reviewed at all stages of a project and updated to ensure that they are as realistic as possible, and that they reflect either improved information on an activity (e.g. subcontractor predicted resources and output data), increased or decreased content, expected or actual resource availability, or changed productivity quotients.

Change in logic

5.7.22   The network logic is as important to the accurate modelling of time as are activity durations, hence any change to logic must be recorded. The review and revision of logic at any stage can improve time and resource usage and overcome problems of resource availability (cost, materials, plant, labour, working space and so on).

5.7.23   The level at which logic is reviewed and revised within a schedule will have varying degrees of impact. Changes to the logic in a Low Density schedule tend to be fundamental, whereas those at Medium Density may affect only the relevant work package, and at High Density may affect only a few activities.

Change in cost profile

5.7.24   Changes to activity costs can be extensively carried out at the project planning stage as part of the scenario planning process in order to maximise the effective use of resources. Costing reviews and changes are likely to be closely interlinked with the overall review of the schedule, and when these take place at the working schedule stage the effects should be carefully analysed, especially where there are contractual issues. (For example, the effect of costing the schedule on interim valuations may have to be considered.)

5.7.25   After any change, it is important to review how costs are allocated, whether by unit rates, total effort or lump sum, by activity, or by resources for the activity, and to check the cost centre alignment against the WBS.

Consequential change in criticality

5.7.26   The results of revisions, especially to durations, logic and resource levelling, will have an effect on float values. However, no attempt should be made to artificially change float values by the use of constraints.

5.7.27   It is inevitable that the critical path will change during the life cycle of the project, and therefore a review is needed in order to determine the effects of change of any part on the whole schedule and its critical path.

5.7.28   The critical path and its activities should 'seem to make common sense' to an experienced reviewer. Consequently, it is important to formally review the revised critical path to ensure that the sequences, logic and durations are appropriate.

5.7.29   It is important to be aware of the chosen software's limitations in relation to its time and/or resource analysis calculation, and consequently the reliability of the calculated critical path indicated by the software.

5.7.30   As with float, the critical path should be permitted to look after itself and be a mathematical calculation of the current schedule. No attempt should be made to manipulate it by the use of constraints or defective logic such as the imposition of negative lags.

5.7.31   Variations between the critical path achieved as a result of review and revision and the preceding critical path will be an expression of the consequences of the changes made to the nature, duration and logic of future activities.

5.7.32   The critical path may consist of more than one string of logic, with other parallel activities involved, and a check should be carried out on all activities with total floats of less than (say) 10% of their durations.

5.7.33   Particular attention should also be given to any path which is predicted to extend beyond any key date, sectional completion date or the completion date.

5.7.34   It is also useful to review the driving activities to all key dates, sectional completion dates or and the completion of discrete contracts and subcontracts.

5.7.35   In reviewing the revised critical path one should consider:

■   logic (including leads and lags)

■   estimated activity durations

■   float

■   unexpired contingencies

■   resource levels

■   constraints

■   method of calculation

5.7.36   On completion of all changes, the critical path should be recalculated and the resultant changes noted and recorded in the planning method statement and all parties notified.

5.8 Change control

5.8.1   For many reasons, intervening events present a special problem in record-keeping, not least of which is the difficulty of keeping abreast of the quantity and frequency of instructed, implied and unintended changes.

5.8.2   Notwithstanding that the contract may be clear and unambiguous as to who bears which risk, at a detailed level the facts may all too often be difficult to label. Accordingly, in the event of any doubt, it is prudent to maintain records of work that might legitimately fall into the class of an intervening event by attributing to it an independent activity coding classification, which stands the work apart from that which falls under the contract. Typically, such content coding might legitimately commence with 'EV' (as an abbreviation of 'event' where that abbreviation has not been used as part of an activity content code for any part of the works).

5.8.3   The recording of change will require capture of the following information:

■   a unique event identifier code;

■   description of the event;

■   originator and/or authoriser;

■   relevant contract clauses providing for extension of time;

■   relevant contract clauses providing for compensation;

■   date upon which the event is instructed/occurred;

■   responsible parties;

■   the activities added, changed or omitted;

■   the labour and plant resources for each added or changed activity;

■   the planned (or, if it has already occurred, the as-built) date and timing of the added or changed activities;

■   the location in which any added work was carried out; and

■   the work flow process adopted in carrying out the change.

5.8.4   There are advantages to be gained from adopting a separate classification for tracking changes that have occurred or have been properly initiated by a party from those that can only be identified as potential changes.

5.8.5   As with information flow and quality control documentation, change management information can be managed by a proprietary DMS software product. However, if these records can be integrated or synchronised with the database used for the progress data, it will be an advantage.

5.8.6   For guidance on the sort of risks often borne by the employer, reference should be made to Appendix 1. However, for information on what risks are actually borne by the employer, reference must be made to the particular contract in question.

Identifying intervening events

5.8.7   Intervening events are unplanned occurrences. They may adversely affect productivity and/or progress and they are often difficult to identify. As a result, it is often the case that the secondary effect of many intervening events (a delay to progress) will be initially identified as a result of the schedule update. The causal event must then be identified, retrospectively, as a result of its effect.

5.8.8   Intervening events can be classified in a number of ways; however, the primary classification is usually by reference to liability. For example, during the construction contract, those that are:

■   at the contractor's risk; and

■   at the employer's risk.

5.8.9   However, for the purpose of management of time, it is important to consider all risks, irrespective of liability and, if disputes are to be avoided, all intervening events and their consequences should be clearly identified and agreed, on a rolling basis, irrespective of liability. It is in everyone's interest to minimise the consequences of an intervening event (which requires proactive time management); if liability is in doubt, it can be established later.

5.8.10   In order to undertake any meaningful analysis of the effect of an intervening event, identification of the date upon which the event itself was initiated, together with the chain of causation arising, will be of the utmost importance. This is because:

■   it will clarify whether events have happened sequentially, in parallel, concurrently, or simply to keep pace with other work;

■   it will assist in distinguishing the effect of one event from that of another;

■   it will determine the calendar date after which an event can possibly have an effect;

■   it may determine the point from which a notice under the contract may be required to be given; and

■   it may determine the time at which the statutory limitation of liability provisions commences.

5.8.11   Whilst the detail of what is required will differ between events of different character, in principle, in relation to every intervening event, the points which must be addressed are those listed above (see Section 5.8.3).

5.8.12   The logic of the intervening event should be set out clearly, together with the activity, or activities, it affects, and the way it affects them. If, for example, a planned activity has actually started when late or revised information is provided, the late information cannot logically inhibit the start of that activity. That will be so, even if the logic of the sequence was planned on the basis that all information was to be provided before the activity in question could start. Under those circumstances, if the late information has any effect at all, it will be because the planned duration of the activity has increased, either:

■   by the time it takes to carry out the additional work in relation to the planned finish of the activity; or

■   because of the delaying effect of the disruption caused by the need to reorganise work as a result of the late arrival of the information.

5.8.13   Apart from changes arising out of the supply of information, there are a number of specific types of intervening events that require special consideration, such as:

■   voluntary and implied variations and other instructed changes;

■   instructions for the expenditure of prime cost and provisional sums;

■   acts or omissions of the employer, or those for whom it is responsible;

■   acts or omissions of third parties;

■   other occurrences; and

■   disruption.

Voluntary and implied variations and other instructed changes

5.8.14   Although the initiation date of an agreed variation is relatively easy to determine, many variations result from changes in design information or from instructions not expressly acknowledged as variations. Typical examples of the latter include extra work arising from:

■   unwarranted rejection of contractually conforming work;

■   unwarranted rejection of contractually compliant submissions;

■   issue of amended documentation to correct a discrepancy in the employer's requirements;

■   issue of amended documentation to correct an error in the bill of quantities, or specifications; and

■   responses to requests for information requiring additional or changed work.

Variations

5.8.15   Most construction contracts will have detailed descriptions of what constitutes a variation and under what circumstances a notification or claim should be made. Additions and omissions are both types of variations.

5.8.16   A variation cannot begin to have an effect upon the contractor's performance until the contractor knows about it and is at least in a position to act upon it. Accordingly, when variations are instructed expressly, or implied, the event should normally be construed to have occurred when the contractor can be shown to have received or known about the instruction. This issue can be critically important, for example, when design changes are made to a part of a BIM model to which the contractor has access, but may not be using the affected part at the time the changes are made.

5.8.17   Contract conditions sometimes specify when instructions are deemed to have been received. For example, the contract may specify that oral instructions are to be confirmed in writing by the contractor within a limited time before they become effective or, if not subsequently rejected within a fixed period of time, they are deemed to take effect on a future date. In the absence of contractual clarity, the parties need an agreed protocol for informing each other of the existence of an instruction that may constitute a change.

Prime cost and provisional sums

5.8.18   With regard to instructions for the expenditure of prime cost and provisional sums, it is good practice for the date (by which an instruction for the expenditure of a provisional sum, a prime cost item, or the nomination of a subcontractor or supplier is required) to be indicated in the working schedule.

5.8.19   Inasmuch as it requires the participation of both the contractor and the nominated party, an instruction for the appointment of a nominated subcontractor or supplier differs in character from other instructions. The content and timing of the nominee's obligations cannot properly be fully ascertained until:

■   the party making the nomination has stipulated their requirements in an invitation to tender;

■ an acceptable tender has been received;

■ an instruction has been issued to the contractor to place the subcontract with the nominee;

■ the contractor has successfully negotiated its requirements with the nominee; and

■ the subcontract is placed.

5.8.20 Accordingly, the initiation date of a delay to nomination should only be taken to be the date upon which the effect of the nomination, if any, can be calculated from the terms of the accepted nominated subcontract. In most cases, this can be expected to be the date upon which the contractor has successfully negotiated its requirements with the nominee that permits a proper prediction of the effect of the instruction.

Employer's acts or omissions

5.8.21 *Health and safety*: Where the contractor is to be excused delay to completion caused by the employer's compliance with, or non-compliance with, occupational health and safety regulations, the employer's obligation is to ensure that the principal designer and the principal contractor under the regulations perform their duties competently. The initiation of the event in such circumstances will be the date of the breach of obligation by the principal designer or principal contractor.

5.8.22 *Late information*: The date upon which a failure to issue any information, drawing details or instructions requested or required under the contract is an employer's time risk event, subject to the terms of the contract. In some contracts, for example, there are two separate provisions:

■ The contract may provide for the employer to determine when information will be provided and this may be set out, for the purposes of tender, in an information release schedule. In such circumstances the event will be initiated on the day after the information was scheduled to be provided in the information release schedule, irrespective of whether it is then needed.

■ Under most forms providing for the contractor's request for information rather than the employer's offer of information, the date of initiation is usually the day after the end of the period during which the information requested should reasonably have been provided.

When information should reasonably have been provided will usually depend upon what is stated in the contract provisions, and if nothing is provided, then an objectively reasonable period should be applied by reference to the published working schedule illustrating the proper commencement and continuance of the work.

5.8.23 *Late response to submission*: The failure to respond in due time to a contractor's submission is initiated not on the date the submission is made, but on the date by which the response was due to be received by the contractor. In many contracts this is stipulated in relation to all submissions, irrespective of their content or importance, as a fixed period of time after the submission is made. In other contracts there may be differing approval periods, depending upon the nature of the submission.

5.8.24 *Response to submission*

■ This can initiate secondary processes, either as a result of the response requiring a variation in the quality or quantity of the works, or in the way the works are required to be carried out (in which case the principles as to when a variation is initiated apply), or as a result of the response properly requiring a resubmission (in which case, for the purposes of managing time, the initial submission can be deemed to be without effect, the submission and approval process then having to recommence from the beginning).

■ Failure to grant access or possession – depending upon the phraseology used in the contract, a failure to give, or deferment of, possession of, access to, or egress from any part of the site is initiated on the day after it was warranted that such possession, access or egress would be given, either by virtue of the contract documents or by subsequent agreement.

■ Alternatively, in the absence of any such specific warranty, the event could be construed to have been initiated after it was reasonably necessary for the contractor to have such possession, access or egress in relation to their published working schedule illustrating the proper commencement and continuance of the work.

5.8.25 *Failure to obtain consents*

■ A failure to obtain any third-party consents necessary for the works is initiated on the day after it was warranted that such permissions would be in place, either by virtue of the contract documents or by subsequent agreement.

■ Alternatively, in the absence of any such specific warranty, the event will be initiated after it was reasonably necessary for the contractor to have such consents in relation to their working schedule illustrating the proper commencement and continuance of the work.

5.8.26 *Suspension of the works*

■ If the works are suspended by an instruction to that effect, the instruction takes effect when the works are suspended and the initiation date is that on which the delay to progress actually commences as a result of the suspension.

■ A power to suspend the carrying out of the works following a failure to make prompt payment is a relatively new addition to the risks borne by the employer in relation to delay arising from the provisions, in the UK, of the Housing Grants Construction and Regeneration Act. This Act entitles the contractor to suspend the performance of the work if it is not paid promptly and in full, in accordance with the conditions of contract. For these purposes, the event is initiated on the date the contractor exercises its right to suspend the performance of the works.[9] Similar provisions may also be incorporated by contract to bind the parties to the same effect where there is no relevant legislation.

5.8.27 *Other occurrences*: In relation to other breaches of contract comprising, for example, any delay, impediment or prevention caused by, or attributable to, the employer, the employer's personnel or the employer's other contractors, the date of initiation of the event is the date the breach is committed.

---

[9] Similar provisions appear in legislation providing for adjudication of disputes in a number of other countries, for example Australian Capital Territory (Building and Construction Industry (Security of Payment) Act 2009); Malaysia (Construction Industry Payment and Adjudication Act 2012); New South Wales (Building and Construction Industry Security of Payment Regulation 2001); New Zealand (Construction Contracts Act 2002); Northern Ireland (Scheme for Construction Contracts in Northern Ireland Regulations (Northern Ireland) 1999); Northern Territory (The Construction Contracts (Security of Payments) Act 2004); Queensland (Building and Construction Industry Payments Act 2004); Scotland (Housing Grants, Construction and Regeneration Act 1996 (Scotland) (Commencement No. 5) Order 1998); Singapore (Building and Construction Industry Security of Payment Act 2004); South Australia (Building and Construction Industry Security of Payment Bill 2009); Victoria (Building and Construction Industry Security of Payment Act 2002); and Western Australia (Construction Contracts Act 2004).

Acts or omissions of third parties

5.8.28   Identifying the initiation date of acts or omissions of third parties is relatively straightforward as they are of no relevance to the contractor unless they actually affect the contractor's works in some way.

5.8.29   To this extent the principles are similar to those on which the timing of a nomination of a subcontractor is construed. Civil commotion, strike or lockouts, for example, are not intervening events at the employer's risk unless and until they affect the progress of the works. In circumstances such as these, the initiation date is that on which the delay to progress actually commences as a result of the event.

Neutral events

5.8.30   Neutral events are those that, under the contract, are the employer's time risk but contractor's cost risk. Often these comprise events which are not really in the control of either party, such as adverse weather, force majeure, labour strikes, and so on. As with the effect of events caused by others, unless they actually affect the progress of the works, they are of no interest in relation to the management of time. The occurrence of these events can sometimes be predicted a short time before they are initiated but for most purposes, the timing and duration of the effect on the progress of the works cannot properly be predicted and will be a matter of record of what actually happened.

Disruption

5.8.31   Disruption results in a loss of efficiency and is usually manifest in increased costs for a given amount of work. It may be caused by a direct delay to progress, or indirectly due to the need to reorganise and redirect work; and it may cause delay to progress. It is often complicated by the fact that disruption to a single activity over a short time-span can be caused by more than one event at the contractor's risk and also others at the employer's risk. However, the overall delay can be retrospectively broken down into its constituent parts by reference to the progress records.

5.8.32   The calculation depends upon what was achieved when the work was not adversely affected by the intervening event that is being investigated. In more complex situations this can be calculated by extrapolating the cumulative resource and productivity data from unaffected progress records of the history of the activity or activities in question and comparing them with the progress records affected by the disruption or, in simple cases, by a comparison between the planned resources and productivity against the actual resources used and productivity achieved.

5.8.33   In summary, depending upon the circumstances, the evidence of lost efficiency is derived from a comparison between either:

■   the planned resources and productivity compared with the resources and productivity actually achieved; or

■   the resources and productivity actually achieved during an undisturbed period compared with the resources and productivity achieved during a disturbed period.

Calculating the effect of intervening events

5.8.34   The effect of an intervening event is calculated as the effect, if any, which the event has on the planned future conduct of the work.

Part 5 - Managing the Dynamic Time Model

5.8.35   Where the event consists of a suspension of the works (in whole or in part) provided that the appropriate calendar is defined, it is satisfactory simply to suspend the working calendar (by inserting non-working time) for the appropriate activities for the appropriate period and recalculating the critical path. This is a technique that is particularly useful for dealing with short suspensions (for example, those caused intermittently by bad weather), which, whilst they may affect the whole of the works, may also, in some circumstances, affect only a few activities, which can be reallocated to a specific calendar for that purpose.

5.8.36   Where a fragnet[10] is used to model the consequences of the event, the effect of the intervening event is calculated by adding the fragnet to the working schedule, making the appropriate logical connection between the fragnet and the affected activity or activities and recalculating the critical path.

5.8.37   The impact of the event is calculated by reference to its effect upon the working schedule at the time of its initiation.[11]

5.8.38   The feature of time impact analysis which distinguishes it from progress monitoring is that whereas the former demonstrates the effect of discrete events on progress and on completion, the latter merely identifies slippage against a target, irrespective of the cause or criticality.

5.8.39   There are two processes by which the time impact analysis method can be used: one which requires the schedule to be updated and impacted to the initiation date of each event sequentially, irrespective of reporting periods; and the other which requires all events within a reporting period to be impacted sequentially upon the last updated schedule. The latter is more common on major projects and is commonly known as the windows analysis process or 'contemporaneous period analysis',[12] each update period being the 'window' in time which is impacted and reported upon.

5.8.40   The process to be followed requires first that the network is updated and rescheduled to identify the calculated effect of progress up to the data date. The events initiated after the data date (but before the next data date) are then impacted in chronological order of initiation date, and the critical path recalculated after each event. This indicates which events will have any effect in relation to the effect of any other and calculates that effect upon the timing of future activities and the critical path. It is a prospective method, which predicts the likely effect of the event on the planned sequence at the time of initiation.

5.8.41   The process is the same irrespective of however many key dates or sectional completion dates are to be considered and whether or not the effect upon any one or more subcontracts or other contractors is to be calculated. However, for ease of interpretation, and applicable to whatever is to be considered (subcontracts, contractors, key dates, sectional completion dates or completion), it is important that the dates to be considered are flagged with milestones, which can be coded and independently organised and filtered for reporting purposes.[13]

5.8.42   Prior to commencing any impacting, the updated schedule should be copied to another; the updated schedule can then be archived and the impacted schedule can be maintained as the operative schedule.

---

[10] A fragnet (fragment of a network) is one or more activities that model the event and its consequences.
[11] See 'Variable baseline target (dynamic)' at Paragraph 5.9.14.
[12] American Association of Cost Engineers International, Recommended Practice No.29R-03 (2009).
[13] See 'Milestone monitoring' at Paragraph 5.9.26.

5.8.43   On the addition of each event, the critical path should be recalculated and recorded and the resultant changes to the critical path noted and recorded in the planning method statement.

5.9   Progress monitoring

5.9.1   Monitoring of progress is the process of identifying, in relation to a target, whether progress achieved has met, exceeded, or failed to meet the target.

5.9.2   Progress monitoring has the useful function of facilitating the detection of trends and facilitating high-level reporting. However, unless change is accommodated by re-baselining, in a major project (in which change is inevitable) it will be difficult, if not impossible, to deduce anything meaningful from a comparison between the base-line and progress achieved.

5.9.3   There are a number of accepted methods of progress monitoring, the most common methods being:

- target baseline schedule (rescheduled compared with baseline)
- jagged line
- count the squares
- milestone
- cash flow
- earned value
- resource

5.9.4   Progress monitoring without rescheduling the critical path (e.g. cash flow, earned value management, jagged line, count the squares, static baseline) should not be adopted as the sole method of managing time in a major project because they have two significant shortcomings:

- they cannot usually distinguish between critical and non-critical activities, or detect shift in criticality; and
- it is generally impossible to predict the effect of a discrete intervening event, simply because they cannot distinguish between the effects of different events occurring over the same time frame, nor can they distinguish between the slip-page caused by events at the contractor's risk and delayed progress caused by those at the employer's risk.

Schedule comparison

5.9.5   This method is not possible with some scheduling software products as it requires the software to illustrate two or more schedules at the same time on an activity-by-activity basis. The sort of display envisaged is illustrated in Figure 36.

5.9.6   This is the simplest and potentially most useful method of monitoring pro-gress and compares what was formerly intended as the construction schedule against what has actually happened. Because it is carried out in discrete steps, it isolates cause and effect and can be used to allocate liability for delay and hence the party responsible for recovery or acceleration.

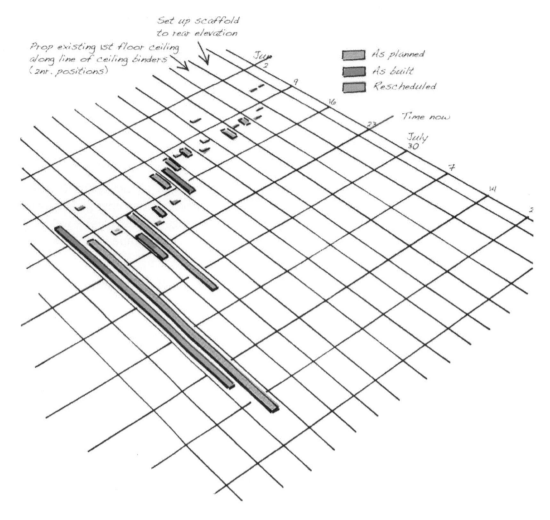

Figure 36 Target schedule comparison.

5.9.7 The results of the comparison are, typically, reported in a bar chart format along with a narrative to explain why the target planned intent was not achieved and why activities or sequences became delayed and prolonged.

5.9.8 It is a useful method for mapping any departure from a previous intent and, if used in conjunction with filtering, can be used to illustrate such things as a shift in timing of subcontract works and change in the critical path.

5.9.9 Because both the target and the recalculated schedule are critical path networks, this method also may provide the necessary information for tracing the cause of departure and identifying what can be done to correct it.

Baseline target schedule (static)

5.9.10 Essentially, the method requires the scheduler to:

■ identify the baseline (i.e. the first) schedule as a target;

■ update, revise, or impact the working schedule with the data to be analysed;

■ reschedule the working schedule to a given data date and recalculate the critical path;

■ save the recalculated schedule;

- compare the recalculated schedule with the original baseline schedule;

- identify the differences.

5.9.11 This method of analysis tends to become inaccurate over time as more instructed changes and other intervening events necessarily affect the progress achieved, while they are not reflected in the baseline target. Thus, the cause of the differences calculated between the recalculated schedule and the target baseline cannot easily be identified, other than intuitively.

5.9.12 When the differences between the recalculated schedule and the baseline are so extreme as to demonstrate that the analysis no longer makes any sense, this method of analysis requires that the status of the project be redefined in a new baseline schedule against which the process can be repeated.

5.9.13 In simple, short-term projects which are not affected by significant changes (and any departure may thus reliably be deduced to be caused by progress-related issues alone), this provides a good basis for schedule analysis. However, on major projects, unless re-baselined to identify the effect of intervening events independently of progress (see Section 5.9.16), using the baseline schedule as a target tends to be an unreliable basis for schedule analysis.

Variable baseline target (dynamic)

5.9.14 If changes have been made to the project requirements or intervening events have occurred, the process requires that the schedule that contains the revised requirements and intervening events that have occurred prior to the last update is called up as the target. By comparing the current state of the schedule with that revised schedule, the effect of progress achieved in the current period can be identified. Only in this way can the effect of the most recent departure from the target caused by an intervening event be identified discretely and independently of the effects of slippage of progress and/or prior intervening events.[14]

Delay caused by a contractor's risk event

5.9.15 In outline, the method requires the scheduler to:

- identify the current saved schedule as the target;

- update the schedule with progress, the effects of which are to be analysed;

- recalculate the critical path of the schedule to a given data date;

- save the recalculated schedule under a new file name;

- compare the updated schedule with the target schedule and identify the differences.

5.9.16 Because the only difference between the target and the schedule being analysed is the addition of progress data, whatever differences are found from the comparison can reliably be identified as having been caused by the progress achieved being different from the progress that was planned. If the difference is a delay to progress that predicts a delay to a key date, sectional completion date or the completion date, the responsibility for recovery will be that of the contractor.

---

[14] For a demonstration of this in use with a varying baseline, see Part 3, 'The dynamic time model' and Appendix 4.

Delay caused by an employer's risk event

5.9.17   Where the purpose of the analysis is to identify the effect of an event at the risk of the employer, rather than the effect of delayed progress, the process is as follows:

■   identify the current saved schedule as the target;

■   impact the schedule by adding the events, the effects of which are to be analysed;

■   reschedule the impacted schedule to a given data date and recalculate the critical path;

■   save the impacted schedule under a new file name;

■   compare the impacted schedule with the target schedule and identify the differences.

5.9.18   Because the only difference between the target and the impacted schedule being analysed is the addition of the event at the employer's risk, whatever differences are found from this comparison can reliably be identified as having been caused by the employer's risk event. If the difference is a delay to progress that predicts a delay to a key date, sectional completion date or the completion date, the responsibility for acceleration to overcome that will be that of the employer.

Jagged line

5.9.19   This method does not compare schedules but identifies the difference between the planned and actual progress on a static baseline schedule printed in the form of a bar chart. A vertical line is drawn, starting at the top of the chart on the data date/progress date extending to the bottom of the page. Where progress actually achieved on an activity departs from that planned, the line deviates from the vertical to a point on the activity bar representing the amount of progress actually achieved. The result is a 'zigzag' or 'jagged line' extending ahead of the data date where progress is ahead of plan, and behind the data date where progress is delayed.

5.9.20   Jagged line progress monitoring does not require a critical path network and is most often used with a bar chart or partially linked network.

5.9.21   Jagged line progress monitoring will not identify what has caused a departure from required progress, and hence any liability for it. Neither will it identify the effect upon the critical path or paths, nor what can reasonably be done to recover from a predicted delay to key dates, sectional completion dates or the completion date.

Count the squares

5.9.22   Progress monitoring by the 'count the squares' (CTS) method[15] is a technique that has been used for many years as a means of assessing progress for the purpose of high-level reporting in simple projects in which the schedule is prepared as a bar chart, but it is of little use in major projects.

5.9.23   The CTS method is a progress monitoring tool in which the information provided is a superficial indication of progress as a whole, rather than specific to a particular activity.

---

[15] See *Project Sponsorship: Planning and Progress Monitoring, Guidance Note No. 7*, The Central Unit on Purchasing, HM Treasury (1986).

5.9.24   In CTS, the work content of each activity is taken to be proportional to its duration and the value of each time unit (usually a day) is the same (unless subjectively 'weighted'). Progress is assessed by counting the number of 'squares' (e.g. days of progress on all activities) planned to be achieved at the status date and comparing them with the number of 'squares' actually achieved. This is a simple process that tends to produce results that are unreliable in all but the simplest of projects.

5.9.25   CTS monitoring will not identify what has caused a departure from required progress, and hence liability for it. Neither will it identify the effect upon the critical path or paths, nor what can reasonably be done to recover from a predicted delay to completion.

Milestone monitoring

5.9.26   A critical path network schedule filtered to illustrate only the milestones is known as a milestone schedule. Typically, milestones may be adopted to mark such things as:

- completion of user requirements and sign-offs;

- design stages and interfaces;

- applications for licences and permissions;

- key dates for 'topping-out', 'weatherproof', or 'power-on', or other significant events in the progress of the work;

- contract requirements;

- payment stages;

- sectional access and/or completion dates;

- the start and finish of a subcontractor's work, or phase of work;

- delivery and removal of key personnel, facilities, equipment, or other important resources;

- handovers and interfaces with key suppliers, statutory authorities and third parties; and

- significant dates for receipt or delivery of approvals and information.

5.9.27   Whilst it is self-evident that a milestone can never be partly achieved, unless it is closely defined, it may be a matter of opinion whether or not achievement has in fact been accomplished by a given date. In order to achieve a common understanding, milestone definitions should be established, setting out what the state of the works must be in order for the milestone to be deemed to have been achieved. The milestone should also be logically connected to the last activity planned to be completed before the milestone can be considered to have been achieved.

5.9.28   Provided they are logically linked to the appropriate predecessor and successor activities, when properly reviewed, revised, updated and impacted, the schedule will reflect the dates upon which the milestones are likely to be achieved.

5.9.29   Where, for visual comparative purposes, a contractually required milestone date is also required to be indicated, this must be identified as a separate milestone pinned to the contractually required date with a mandatory date constraint, but it must not be linked to the network. This combination is illustrated in Figure 28.[16]

---

[16] See also 'Inflexible constraints' at Paragraph 4.27.3.9 et seq.

5.9.30   Milestone monitoring will not identify what has caused a departure from required progress, the effect upon the critical path or paths, nor what can reasonably be done to recover from a predicted delay to completion. Milestone monitoring is thus usually used for high-level reporting in conjunction with other monitoring processes to identify predicted delay to key dates, sectional completion dates or the completion date, and to plan recovery and/or acceleration.

Cash-flow monitoring

5.9.31   This a financial performance monitoring technique that relies entirely upon the analysis of cash flow. Essentially, the method requires:

■   a planned cumulative cash-flow forecast; and

■   the cumulative certified amount/or costs to date.

5.9.32   Monitoring is achieved by a comparison between the two costs, usually by cumulative graph against data dates. The departure between the lines of the graph, if any, will indicate by reference to the data dates approximately when the departure occurred (but not the cause of it). By extrapolation it may also be used to predict the total ultimate cost of the works

5.9.33   The major difficulty in cash-flow monitoring is the problem of determining the precise value of money expended at the data date. Financial accounting systems that are dependent on receipt of an approved invoice, for example, frequently take 4 to 6 weeks to receive and process all of the payment information. However, this difficulty can be avoided where a properly processed dynamic time model is used to calculate any revisions to the planned cost arising out of changed content, sequence or method, and the updated schedule is used to calculate the actual cost.

5.9.34   If a different work content or sequence of working from that planned is adopted, a meaningful comparison between planned cost and actual cost will be difficult, if not impossible.

5.9.35   Cash-flow monitoring will not identify what has caused a departure from the planned cash flow forecast, the effect upon the critical path or paths, nor what can reasonably be done to recover from a predicted delay in achieving the total planned expenditure. Cash-flow monitoring is thus usually used in conjunction with other monitoring processes to identify predicted delay to key dates, sectional completion dates or the completion date, and to plan recovery and/or acceleration.

Earned-value management

5.9.36   Earned-value management (EVM)[17] is another financial performance monitoring technique based on the comparison between planned value and earned value. It is a more sophisticated method of financial monitoring than cash-flow monitoring as it attributes actual rather than notional values to the activities, either by reference to the certified value or the value of labour, materials and plant resources, or both. This method of financial performance modelling does not take account of costs not attributable to a work package such as general site-related plant, but it can give a useful illustration of the value of work achieved in relation to the value of work planned to be achieved in a given time period.

5.9.37   EVM is established by allocating a budget to the plan (usually this is in terms of money and man days per activity). From that data a cumulative expenditure

---

[17] For a detailed explanation of EVM, see BS6079-1:2002; AS4817:2006; ANSI748B.

graph can be drawn for the project overall and for the elements within the WBS, known as the planned value.

5.9.38   Once this has been done, a copy of the budget plan is made to act as a target baseline. In order for EVM to give meaningful results, it is vital that the budget target is maintained with change as it occurs. The project change control process will govern the process of change, but this must involve both contract change (change in scope) and preferential change (change in methodology; development from Low to Medium to High Density schedules).

5.9.39   As with all other financial monitoring processes, one of the challenges of earned-value management is that of determining the precise value expended at the data date. Actual values are collected and collated at an appropriate level. This will be at a level suitable for management review, and will usually be at a higher level than that of the activities in the working schedule.

5.9.40   Again, as with all other financial monitoring processes, if a different work content or sequence of working from that planned is adopted, a meaningful comparison between planned value, earned value and actual cost will be difficult, if not impossible. However, this difficulty can be avoided where a properly processed dynamic time model is used to calculate any revisions to the planned value arising out of changed content, sequence or method, and the updated working schedule is used to calculate the earned value.

5.9.41   Accordingly, for the purposes of time management, the dynamic time model[18] should be used to identify the effect of changes in content, sequence and method. The critical path is recalculated to allow the working schedule to remain a dynamic model, detailing the logical sequence and prompting what can reasonably be done to recover from any predicted delay to key dates, sectional completion dates or the completion date.

5.9.42   The planned value, earned value and actual cost can be compared, using visual charts and key performance indicators (KPIs). The two most useful KPIs and the basis for calculations of future performance are the schedule performance index ('how are we doing against planned expenditure?') and cost performance index ('are we efficient?').

$$Cost\ performance\ index = \frac{Earned\ value}{Actual\ cost\ to\ date}$$

$$Schedule\ performance\ index = \frac{Earned\ value}{Planned\ value}$$

5.9.43   The benefits of EVM are:

■   identification of the value of work that has been achieved against the planned value of work, and what it has cost to reach that level of achievement;

■   whether the work achieved has been achieved efficiently (i.e. represents good value for money);

■   whether the project is likely to finish on budget; and

■   whether it is likely to take longer than planned to achieve completion of the project.

5.9.44   EVM monitoring alone will not identify what has caused a departure from required progress, the effect upon the critical path or paths, nor what can reasonably

---

[18]   See Part 3, 'The dynamic time model' and Appendix 4.

be done to recover from a predicted delay to completion. EVM monitoring should thus be used in conjunction with the dynamic time model to identify relevant values and in order to predict delay to key dates, sectional completion dates or the completion date, and to plan recovery and/or acceleration.

Resource monitoring

5.9.45   Resource monitoring can be carried out against labour, plant and/or materials. The principles are the same as in other methods of progress monitoring, in that there must be a meaningful target (of like content and sequence) against which current resources can be compared.

5.9.46   With resources, the comparison can be project-wide or more discretely focused upon a particular zone, element or subcontractor, as required. The only prerequisite is a meaningful target against which actual resources employed can be measured against the resources planned to be employed.

5.9.47   Data will usually be contained in daily allocation data input sheets, which can be summarised, and graphs can be produced to show actual resources used compared with those planned and, by extrapolation, predict consequences.

5.9.48   Resource monitoring will not identify what has caused a departure from required progress, the effect upon the critical path or paths, nor what can reasonably be done to recover from a predicted delay to key dates, sectional completion dates or to the completion date. Resource monitoring is thus usually used in conjunction with other monitoring processes to identify predicted delay to key dates, sectional completion dates or the completion date, and to plan recovery and/or acceleration.

Building information modelling

5.9.49   Different 4D modelling software packages provide options for different methods of progress monitoring, which can be simulated in the virtual 3D model. For example, in certain software it is possible to record as-planned and as-built information, with separate 3D virtual models displayed at the same time. This allows the possibility of visually comparing progress in a virtual environment and navigating the model.

5.10   Acceleration and recovery

5.10.1   Irrespective of the causes, or whose responsibility it is, at some point in every major project it will be necessary to consider recovery or acceleration in order to overcome the effects of a delay to progress.

5.10.2   The strategy for dealing with recovery and acceleration should be a matter of policy and set down in the planning method statement.[19] In principle, the options are usually changes to duration, sequence, method or resources, and may include one or more of the following planned processes:

- design
- standards of workmanship or materials
- method of working
- quantity of work

---

[19]   See 'Planning method statement strategy' at Section 2.1 and 'Planning method statement' at Section 4.3.

- sequence (changes in interfaces/logical dependencies)

- resources

- working time

- contingencies

- information flow

5.10.3   In any particular case, whichever method is to be adopted, its appropriate application will depend upon a number of factors including:

- At whose risk is the delay to progress?

- What portion of the works has it affected?

- What trades or activities has it affected?

- How much time is to be recovered?

- On what path or paths is that time to be recovered?

- What are the activities on those paths?

- Can different methods be applied to different activities and paths?

- What cost is likely to be incurred?

- How effective is the recovery likely to be?

- What are the risks in implementing the selected method or methods?

5.10.4   The priority to be given to the various choices will largely depend upon when, in relation to the plan, the recovery or acceleration is to be attempted and the degree of time to be made up. Whatever strategy is adopted, the answers to the above issues, and hence the reasoning for them and justification, should be clearly set down in the planning method statement consistent with the reviewed and revised working schedule to which the strategy is to be applied.

5.10.5   It should always be borne in mind that, whilst contingencies will be readily identifiable as part of the schedule activities and logic, they should be preserved as long as possible consistent with the current risk appraisal. If contingencies are absorbed in the early stages of a project, the likelihood of completion on time will be reduced. This is simply because, as the project proceeds, the number of alternative methods available to make up time will naturally reduce.[20]

5.10.6   Accordingly, contingencies should be reviewed and revised in the light of the associated status of the risks and the working schedule. When reviewing the contingencies, review and revise time allowances for those activities concerning contractors' and employer's contingencies independently, depending upon the party responsible for the effects of the intervening event which caused delayed progress, the effects of which are to be recovered.

5.10.7   Risk for which a contingency is provided, and which then matures, will be accommodated in the schedule as either one activity, or a series of activities added to it, which then replace (in whole or in part) that contingency.

5.10.8   Where it is evident that the anticipated risk has not matured to the extent provided for, or at all, the contingency will be reduced or omitted entirely and replaced by the as-built series of activities, as they actually occurred.

---

[20]   See 'Time risk management strategy' at Section 2.9.

5.10.9   Redundant contingency should be either added to later contingencies (as a result of reappraisal of enhanced risk) or omitted entirely.

5.10.10   Prior to commencing the planning of any recovery or acceleration, the schedule should be copied to another so that, prior to revision, the schedule can be archived and the revised schedule can then be maintained as the operative schedule.

5.10.11   During revision, an audit trail should be maintained of the changes made, and the reasons for them, recorded in a revision of the planning method statement. Every revision should be carefully assessed to ensure the proposed acceleration can reasonably be achieved and any risk incidental to it is properly accounted for.

5.10.12   On completion of all revisions, the critical paths should be recalculated and recorded and the resultant changes to the critical paths noted and recorded in the planning method statement.

6 Communicating the dynamic time model

6.1 Introduction

6.1.1   Once the planning and scheduling of the work is complete, the primary purpose of the dynamic time model is to facilitate the management and direction of the works that are to take place. This is achieved through communicating the sequences, methods, required outputs and the dates by which the work needs to be undertaken.

6.1.2   Within a BIM environment, the coordinated time and design information in a 4D model can make a significant contribution in communicating the time model and assisting key stakeholders in making informed decisions.

6.1.3   Communicating the information held in a 4D model is not limited to those with the relevant modelling software. 4D models can be exported as both videos and static images. Some 4D modelling software providers have also created viewers, permitting electronic navigation of the model without the permissions necessary for editing.

6.1.4   Allied to the primary purpose, the time model is the management of events, including unforeseen events and changes to the requirements of the project.

6.1.5   The time model is also used for performance management of the project. This is the management of construction performance through the measurement of progress and analysis of outputs in order to inform the instigation of any mitigating actions that are necessary.

6.1.6   Finally, during the project, data is collected that can inform improvements not only in the current project, but in the estimation and planning of future projects. This is known as benchmarking data.

6.2 Proactive communication: promoting the plan

6.2.1   Effective communication takes a number of forms and must be appropriately tailored to suit each audience. Opportunities to promote the plan can be verbal: perhaps at planning workshops and other meetings; or visual, by using summary schedules and alternative means of display.

6.2.2   Schedule communication takes two distinct forms. First there is the preparation and distribution of regular reports containing routine information for the management of the works, each report containing specific information extracted and at an appropriate level of summarisation to best suit the requirements of the recipient.

Guide to Good Practice in the Management of Time in Major Projects: Dynamic Time Modelling,
Second Edition. The Chartered Institute of Building.
© 2018 John Wiley & Sons Ltd. Published 2018 by John Wiley & Sons Ltd.

The second form comprises ad-hoc communications created to inform a particular need or issue. The methods of communication can include:

- planning presentations, workshops and discussions;

- review meetings, especially short-term planning meetings (for the High Density part of the schedule);

- use of summary schedules, prominently displayed around the site offices;

- videos and static pictures from 3D and 4D models;

- various tabulations and extracts of information;

- text-based summaries and reports;

- line-of-balance diagrams;

- time chainage diagrams;

- key milestone schedules celebrating successes and displaying future targets;

- time-based logistics plans; and

- key quantity progress graphs (S-curves).

6.2.3   Where the time model is conceived as a 4D model in a BIM environment, exports from the native 4D model can be imported into other video and photographic editing software packages for enhancement. However, care should be taken to make sure the result does not become misleading by the obstruction or removal of relevant parts of the model by the use, for example, of extreme or unusual camera angles. All presentations and images should be produced with an understanding of the message they are intended to convey.

6.3   Reactive communication: reporting

6.3.1   Reporting is an extremely powerful tool that enables key stakeholders to make informed decisions, with regard to time management, in an efficient and effective way. Good reporting ensures that all parties have the same perspective on the status of the project.

6.3.2   Reporting is the process of disseminating the required information to those who want it, need it, or should be informed. Necessarily, it involves communication of a message which, as in any other form of communication, must be received and understood if it is to be acted upon.

6.3.3   A well-structured report enables the users to extract key information pertinent to their role, but no single report can meet the needs of all of the project's stakeholders effectively.[1] Effective reporting requires striking a balance between the simplicity of only producing one or two reports (that require the recipient to find the information they need from within the data provided) and the challenge of producing multiple boutique reports for each user. With modern web-based communications a balance also needs to be struck between the need to 'push' information out to recipients and making data available for users to 'pull' the information they need, as required. BIM is increasingly moving towards a 'pull' approach to making information available and usable.

6.3.4   Whether the communication is required to be delivered following the occurrence of an intervening event, or regularly (irrespective of events), or occasionally

---

[1] See also 'Schedule communication' at Section 4.12.

(depending upon the nature of the data) determines the significance of the report. What the contract requires in that regard should be identified in relation to the purpose of the report and the parties to whom it is to be directed.

6.3.5   Understanding and conforming to the contract requirements is important because, under some forms of contract, anything issued outside the required time, or not to all the specified parties, may have the contractual effect of shifting the liability for the consequences of certain events from the contractor to the employer, or vice versa.

6.3.6   The form in which the report is to be issued may be stipulated under the contract, but in most circumstances all that is usually required is that it is unambiguously issued on a particular date to a particular party. For those purposes, most reports are issued as documents in one form or another and, if issued orally, are confirmed in writing.

6.3.7   However, whilst being issued as a document infers that it will be in writing, for the purposes of time management, because the facts are usually complex, figures, bar charts, graphs, histograms and fragnets may also be usefully employed in the communication, depending upon the subject matter and target.

6.3.8   At management board level, videos, photographs, simple figures, charts and graphs are extremely useful in communicating complex data. Animations and videos exported from a 4D model can also be extremely useful in communicating complex procedures and sequences. At site level, a more detailed analysis (to enable problem-solving) is often significantly more useful.

6.4   Report types

6.4.1   Reports can be directed to two different types of recipients, external and internal:

- external recipients include the employer, design consultants, financiers and such parties as permitting and licensing agencies;

- internal recipients will include the contractor, construction manager, design consultants, site superintendents, contract administrator, subcontractors and suppliers.

6.4.2   The reports generally fall into the following three categories:

- contractual

- managerial

- informational

Contractual notice

6.4.2.1   Most forms of contract require the contractor to give notice of the occurrence of a notifiable risk or event that is causing, has caused, or is likely to cause a delay to the contract, to the party responsible for administering the contract (the contract administrator) within a limited period following the occurrence. For example, finding ground conditions that are not as described in the contract, which may cause delay, will typically require notification within 7 or 14 days of the discovery, notwithstanding that the full consequences of the different conditions may take several months to be fully defined.

6.4.2.2 For the precise requirements of the notices to be provided and the information to be included, reference must be made to the contract in question. However, in general, where time management is concerned, whereas the 'notice' itself may require little more than the formality of advice that a specified event has occurred and that it is causing, likely to cause, or has caused delay, much more is needed for management of the risk.

6.4.2.3 Because the requirements of the initial notice are often fulfilled by a document containing very little useful information, most forms of contract also call for supplementary information to be provided later, defining the extent of delay to progress and, as a consequence, the predicted extent of delay to sectional completion dates or the completion date.[2] Usually this has to be provided within a further limited period after the initial notice has been served and must also comply with the contract requirements.

6.4.2.4 In addition to any contractual requirements, the supplementary information should contain all of the information necessary to identify:

- the nature of the event and its unique identifier;

- the contractual clauses under which the notice is required;

- a synopsis of the event and the timing of the constituent activities;

- the planned activity or activities affected and the manner of the effect;

- the likely consequential effect on the key dates sectional completion dates and the completion date, if any;

- the likely consequential effect on any other contractors and subcontractors, if any; and

- the file reference of the impacted schedule.

6.4.2.5 Communication is usually enhanced if the textual information in the above items is supplemented by relevant diagrams, such as a picture of the fragnet used to impact the event, together with a trace of the critical paths from the event to the relevant completion dates on the impacted schedule. The information supplied needs to both fulfil the requirements of the contract (to make the notice valid under the contract) and, equally importantly, fully explain the event and its consequences to the contract administrator and others (such as the employer and its consultants), so that the notice and its consequences can be properly managed.

6.4.2.6 In the example of a notice of delay in Appendix 8, the event description is: 'Instructions to resolve a discrepancy between design information'. It is expressed that way because that is the phraseology used in that particular contract to describe the event which entitles the contractor to an extension of time and compensation. The effect, 'delay to the MOF Jetty',[3] is not the event. That is the direct effect of the event.

6.4.2.7 For clarity, the direct effect should be clearly differentiated from the causative event giving rise to it and, if relevant, the consequential effect on any contractors, subcontractors, key dates, sectional completion dates, or the completion date flowing from the direct effect.

---

[2] See, for example, the sample notice of delay in Appendix 8.

[3] A Materials Offloading Facility (MOF) Jetty is used in connection with marine installations for the transport of LNG and LPG.

Managerial reports

6.4.2.8   The planning method statement should include a statement of the contractor's intended reporting protocol, identifying those concerned with the content, form and detail of the various managerial reports that are to be generated (and how the various coding systems will facilitate efficient extraction of the information). Where this type of report is required by the contract, the contract may also set out specific requirements for the report's content and format, or the need for agreement of format and content with the recipient.[4]

6.4.2.9   The purpose of managerial reports, as the name suggests, is to facilitate the effective management of the project. In the Guide we are particularly concerned with the management of time, and for that purpose the managerial reports are usually referred to as 'progress reports'.

6.4.2.10   Essentially, for the purposes of time management, the progress report will contain information set out under the following headings:

- executive summary;

- progress achieved in the last reporting period;

- progress expected in the next reporting period; and

- planned progress in the medium-term and long-term future.

Executive summary

6.4.2.11   This should provide a high-level summary of the project status and key issues affecting the project, including the effect upon:

- key dates;

- other contractors;

- subcontractors;

- employer's deliverables;

- sectional completion dates;

- the completion date;

- work trends;

- risk; and

- key decisions required to be made by the executive.

6.4.2.12   Progress in the last reporting period:

- what was the High Density part of the schedule?

- what actually occurred and when?

- any identified variance between planned and actual performance?

- what caused each difference?

- what is the likely effect of any difference upon:

  ○ key dates?

  ○ other contractors?

---

4 See 'Planning method statement strategy' at Section 2.1 and 'Planning method statement' at Section 4.3.

◯   subcontractors?

◯   employer's deliverables?

◯   contingencies?

◯   sectional completion dates?

◯   the completion date?

6.4.2.13   Progress in the next reporting period:

■   what is the High Density schedule for the next three months?

■   what is in the High Density schedule for the next reporting period?

■   which contractors and subcontractors are involved in the work?

■   what planned resources and productivity are assumed?

■   what employer's deliverables will be required and when?

■   what (if any) risks may occur, or become greater, or lesser?

■   what is the strategy for dealing with those risks?

■   which contingencies will require adjustment and to what extent?

■   what recovery or acceleration (if any) is planned and how is it to be implemented?

6.4.2.14   Planned progress in the medium-term and long-term future:

■   Progress reports will usually be required to be accompanied by reports upon contractual key dates, sectional completion dates and the completion date; each should be treated as a separate critical path and reported upon independently of the other. The relevant issues are:

    ◯   what information has been received to affect the Medium Density and/or Low Density schedule?

    ◯   what recovery or acceleration will be required (if any)?

    ◯   which contractors' contingencies will require adjustment and to what extent?

    ◯   what (if any) risks may occur?

    ◯   what is the strategy for dealing with those risks?

    ◯   which employer's contingencies will require adjustment and to what extent?

    ◯   in summary, what recovery or acceleration (if any) is planned and how is it to be implemented?

■   Where the time model is conceived as a 4D model in a BIM environment, it is important that in schedule review meetings the appropriate software and hardware are made available for viewing the 4D model. For example, more than one adequately sized screen may be necessary to enable all participants to comfortably view the model. Otherwise, the benefit of 4D modelling to facilitate engagement may be limited.

Informational reports

6.4.2.15   This type of report may be required by various stakeholders, in which case the specific requirements will be defined in the specification. Based on this requirement, the planning method statement should include the reporting require-ments, and the content, form and detail of the various informational reports that will be generated (and how the various coding systems facilitate efficient extraction of the information). In general, where time management is concerned, informational reports may be required for a range of purposes, including financial control, quality control and testing, health and safety provisions and accidents, data flow and drawing release, requests for information, change control, and so on. The design of each report will depend upon the type of data to be transferred and what is required to transfer the information effectively.

6.5   Reporting formats

6.5.1   The best format to use for a report or a presentation depends upon for whom the report is being prepared, the information they need and their preferred way of receiving information. A different process and output will be required depending upon whether the report is needed for strategic or operational purposes and whether the recipient is to make a decision, give advice, monitor progress or simply receive the information. However, for most purposes, it will be helpful to have an executive sum-mary at the beginning of any report, highlighting the important points with a cross-ref-erence to the detail. In that way, even the busiest of people can quickly grasp the principles of what is being said and easily drill down to the detail, where they need to.

6.5.2   A good general principle to keep in mind when thinking about the best way to present information is that, at site level, reports will usually be required with precision and detail but limited to a relatively short 'look-ahead' period referable to the High Density part of the schedule; whereas at the executive level, overall summaries with trends and exceptions tend to be more relevant. In all cases the key determinant of an effective report is presenting the information the recipient needs to perform their role on the project, in the easiest format for them to access, understand and use.

6.5.3   Whenever tables of data are necessary, including a visual representation of the information should always be considered as a method of summarising the effect of the data (together with an explanation of the diagram). Useful data visualisation tools include:

■   histogram

■   pie chart

■   graph

■   PDM network

■   ADM network

■   bar chart

■   linked bar-chart network

6.6   Feedback and benchmarking

6.6.1   If the principles of the Guide are followed, every project will provide a plethora of useful data that can be used to improve the performance of both the current pro-ject and future projects.

6.6.2   The data will have been captured throughout the life of a project by refer-ence to the progress records database, the updated schedules and the impacted schedules. This information can be used to identify any discrepancy between the observed state and the desired standard in any particular case. Only after an issue has been described is it possible to instigate process improvements.

6.6.3   There are two aspects to benchmarking. On the one hand, the process will establish achieved norms of productivity and activity durations for common and project-specific working conditions, work types, trades, resources and so on, and on the other hand, for common data it will provide some degree of comparison between the time model performance and industry best practices or standards.

6.6.4   The first step in any analysis of performance data is to identify what is to be analysed. This should be done before construction begins so that the planned work type can be identified for tracking to create the data needed for analysis. The more closely this can be defined, the more easily the required analysis can be performed and the more useful the data will be.

6.6.5   For example, in order to identify the normal productivity rate for a given work type, you will normally need to adopt the following procedure.

6.6.5.1   Identify the various activities comprising the work and examine the relationship between them. If there was a reasonable degree of continuity between relevant activities, then the data can be taken together to represent the performance of that work type from beginning to end. If not, then the analysis will also provide information on the degree of productivity lost as a result of intermittent working (if any) by comparison between the best continuity achieved and that achieved during broken periods. Care is needed to ensure the absence of (or quarantined effect of) any other affecting events.

6.6.5.2   Identify the resources used throughout the various activities and establish whether there is any significant fluctuation between them in continuous and discontinuous operating conditions. If the resources were constant, then whatever fluctuation in productivity is apparent will have been caused by something other than the quantity of resources. The effect of any known event, such as the effect of a learning curve, or an intervening event, can then be isolated and filtered out and the remainder will represent the average normal achieved productivity for that work type. Where there are remaining fluctuations, the best achievable productivity and average productivity can be established relatively easily.

6.6.5.3   Where there are fluctuating resources, then a separate analysis of achievable productivity should be made against each combination of resources to establish the effect of the different combinations on the average and best productivity in normal conditions.

6.6.5.4   Where commonly occurring events have interfered with productivity, data for the effect of those on the selected work type can also be established on differing combinations of resources.

6.6.5.5   Tabulate the results and the criteria adopted for identification of each separate analysis. Depending upon the activity type, the productivity per unit of resource (machine/person/crew) will be described as a quantity of work against a fixed time period, for example, cubic metres per day or linear metres per hour.

6.6.6   Knowing the effect of different conditions on performance, and the circumstances under which the work was carried out, are essential if the analysis of past performance is going to provide useful information to assist in planning future works. To make the best use of such records, the data should be kept in a database, which can be searched, organised and filtered for future use. Typically, in relation to any search, the data retrievable should consist of values in at least the following data fields:

- job name
- job type
- country and region
- date started, completed and construction period
- design team
- project management team
- construction management team
- activity type and whether it is a common activity, or with project-specific difficulties
- activity duration of which data is measured
- best working conditions' characteristics
- best working weather conditions
- best productivity achieved
- average uninterrupted productivity achieved
- average working conditions' characteristics
- most productive resource combination
- average resource combination
- the effect on productivity of specific events (by type, e.g. multiple reissues of drawings)

6.6.7   For some data, such as working conditions, a pick list of possibilities might be established to restrict data input to a defined range of values for sorting, filtering and retrieval. Once the data has been accumulated, then comparison can be made with other known data to establish whether there is anything inherently out of the ordinary. For example, comparison with common activity types can be made against data in trade manuals.

6.6.8   Benchmarking requires a high degree of analytical capability. It can be a time-consuming and costly exercise to complete by those unfamiliar with the research necessary to arrive at meaningful data, and unless carried out to a high standard, against a quality-assured process, the results can be rendered useless. Accessing the required specialist knowledge and capabilities requires either an experienced in-house department, or alternatively consideration should be given to engaging external consultants or a collaborative organisation experienced in the process.

APPENDIX 1 Time risks that may be borne by the employer

Introduction

Subject to the terms of the particular contract, if a risk that is accepted by the employer materialises to cause disruption and/or delay to progress, the contractor may be entitled to compensation for a delay to progress; and if the delay to progress adversely affects an activity on a critical path to a sectional completion date and/or the completion date, the contractor may also be entitled to:

■ an extension of time to complete; or

■ an extension of time to complete and also compensation for prolongation arising out of the delay to progress.

Risks that are within the employer's control

The risks which are usually borne by the employer under standard form contracts are:

An instruction given by or on behalf of the employer to:

■ execute a variation;

■ expend a prime cost or provisional sum;

■ change the employer's design requirements;

■ change the health and safety plan;

■ open up and/or investigate any work, goods, materials to search for defective work, goods, materials where no defect was ultimately discovered;

■ remedy any factual error in information contained in the contract documents;

■ remedy the failure of a nomination;

■ rectify any error in setting-out caused by incorrect data supplied by the design team;

■ repair or replace the loss of, or damage to, the works caused by an insured risk;

■ correct any unstated departure from the adopted method of measurement;

■ correct any error, omission, inconsistency or ambiguity in a model or other information caused by a defect in software and/or hardware specified by the employer;

■ make any tests not described in the contract in sufficient detail for the contractor to have allowed for it in his tender, or defer the specified tests;

■ correct any impossibility or illegality;

■ ensure conformity with any Act of Parliament, regulation or by-law;

Guide to Good Practice in the Management of Time in Major Projects: Dynamic Time Modelling,
Second Edition. The Chartered Institute of Building.
© 2018 John Wiley & Sons Ltd. Published 2018 by John Wiley & Sons Ltd.

■ investigate and report upon the practicality of alternative value engineering proposals;

■ overcome unforeseen physical conditions or artificial obstructions;

■ conform to any condition or limitation in any licence obtained after the award of the contract;

■ dispose of human remains, fossils, coins, articles of value, or antiquity and structures, or other remains, or things of geological or archaeological interest; and

■ suspend the carrying out of the works.

Errors and omissions by the employer or any person or other contractor engaged by the employer:

■ a failure to make available to the contractor the necessary parts of the site on or before the date stated;

■ a failure to provide a response to a request within a reasonable time;

■ an error in any model or other information provided by the employer upon which the contractor is permitted to rely;

■ an impediment, prevention or default, whether by act or omission, of the employer or any person or other contractor engaged by the employer;

■ the execution of a measured quantity of work that differs significantly from any approximate quantities stated in the bill of quantities prepared by, or under the direction of the employer;

■ any failure to consent in due time to a contractor's submission;

■ the application of unreasonable conditions to or rejection of consent to a contractor's conforming submission;

■ interference caused by any other properly authorised authority, or utility, or statutory undertaker in connection with their statutory obligations, or failure to do work;

■ interference caused by an act or omission of a nominated subcontractor and/or nominated supplier;

■ a failure to provide anything which the employer is to provide;

■ the wrongful withholding of consent to the expulsion of a nominated subcontractor, pursuant to any forfeiture clause or rescission of the subcontract;

■ any suspension by the contractor of the performance of his obligations for a failure to make payment;

■ any delay, impediment, prevention or default by the employer or any person or other contractor engaged by the employer.

Risks that are not normally within the employer's or contractor's control

Suspension of the works, any part of the works or disruption as a result of:

■ a failure of a utility company or statutory authority to carry out required work;

■ an order, or other act, or a failure to act in accordance with the law of a court or statutory authority;

- a strike or lockout affecting the suspended part of the works;

- civil commotion, or the use or threat of terrorism dealing with such an event or threat;

- discovery of geological or hydrological condition or man-made or other obstruction, void, conditions pollutant or contaminant;

- discovery of human remains, fossils, coins, articles of value, or antiquity and structures, or other remains, or things of geological or archaeological interest;

- the occurrence of adverse weather;

- the effect on labour, materials or goods of government acts or controls;

- the inability of the contractor to obtain labour, materials or goods, which could not have been foreseen at the date of tender;

- any event beyond the contractor's control;

- any other event outside the control of both the employer and the contractor.

Appendices

APPENDIX 2   Case studies in strategic planning

Introduction

This Appendix illustrates the importance of taking into consideration, at the strategic planning stage, the employer's time risks under the contract and the need for time contingencies for delay in planning to achieve completion on time. It also highlights the need to take into consideration, in budget planning, the cost contingencies for delay-related cost claims which will arise out of the employer's time and cost risks under the contract.

The case studies have been selected to demonstrate that the same types of issues arise in major projects irrespective of the industry, culture or geographical area in which the project is carried out. They concern:

1.   a pipeline in South America
2.   a railway in Australia
3.   administrative offices in Europe
4.   a road in Asia

Necessarily, these case studies have been selected from experience of many projects over many years. None describe the circumstances of any one project, and in each case some facts have been changed to make the illustration simpler and easier to understand. Accordingly, any similarity between these examples and any real project is unintentional.

Case study 1: Pipeline in South America

Project outline

The project involved cutting a pipeline approximately 150 km long, through a forest of great ecological and archaeological importance, containing small village populations. For this section of the route, the employer wanted the cheapest contract cost in the shortest time with the least possible damage to the forest. Special planning considerations included:

■   At one end of the route was a tributary to the Amazon river and at the other the Andes mountains.

■   The right of way was not to exceed 25 m wide. Anything wider than that was to be made good at the contractor's risk.

■   The wet season was expected to extend from October to March, when no work could be carried out at all. This was at the employer's risk.

■   Rain could be expected on odd occasions in the dry season. Rain would prevent work being carried out. This was at the contractor's risk.

■   If the pipeline infringed on an occupied settlement, the employer's compensation team would negotiate compensation with the tribal chief and/or the relevant

Guide to Good Practice in the Management of Time in Major Projects: Dynamic Time Modelling,
Second Edition. The Chartered Institute of Building.
© 2018 John Wiley & Sons Ltd. Published 2018 by John Wiley & Sons Ltd.

Appendices

land owner, before clearance could commence. An average of four weeks' delay was expected per stoppage. This was at the employer's risk.

■ If archaeological remains were discovered, work would stop while the State's archaeological team investigated the findings. An average of six weeks' delay per stoppage could be expected. This was at the employer's risk.

Contract

A project-specific design-and-build engineering contract had been drafted on the employer's behalf. This required the employer to grant an extension of time and to pay compensation for prolongation for delay caused by those risks which were specifically stated to be at the employer's risk. All other risks were to be borne by the contractor.

The strategic plan

Delays to progress could be expected as a result of bad weather, archaeological findings, settlements and diversions and, because of the prohibition on working in the wet season, a few short delays to progress during the working season could easily delay completion of the work into the subsequent wet season and amount to a whole year's delay in project completion.

Each base camp (to which materials could be delivered and where workforce and equipment could be housed) could be expected to be an expensive temporary construction involving site clearance, all services, security, accommodation, storage facilities, roads and so on, and clearance and restoration on completion. The contractor wanted to minimise costs in order to appear most competitive at tender stage and decided on only two base camps, which were relatively easy to access for deliveries: one at the foot of the Andes and the other at the other end of the right of way, at the head of the river.

The planned scheme thus involved two parallel paths achieved by setting up base camps at those locations and driving the pipeline from both ends towards the middle.

The challenge

This was a practical plan on its face. However, what was not planned for was the fact that, at 150 km long, the mid-point was 75 km from either base camp: a point where it would take 4 hrs to reach and 4 hrs from which to return out of a 10-hr working day. Thus, if either leg was delayed, it would not be long before the other leg had progressed as far as it could go (as a result of travel time consuming the whole working day) and the whole line would come to a standstill.

The leg starting at the head of the river suffered significant delays, as a result of factors largely at the employer's risk, which prevented delivery of materials to the camp by river. The contractor attempted to reduce the effect by bringing materials and labour to the head of the delayed route by helicopter. The project finished 18 months late and many millions of dollars over budget. Claims ensued.

A more effective planning solution

Analysis of the possible combination of the various cost and time factors carried out using iThink[1] modelling and system simulation software revealed that the cost of putting in a third camp at the mid-point was not significantly different from the

---

[1] 'iThink' is a product of isee systems.

prolongation costs that could have been expected to have been suffered, purely as a result of those delays to progress which, from inception, could have been expected. Except in the extremes of absolutely no delay occurring at all (in which case the third camp would have been redundant) or absolutely everything going wrong (in which case the project would always have been badly delayed), the introduction of a third base camp at the mid-point had the following advantages:

- it significantly reduced the non-productive travelling time to all work-faces;

- even when delay stopped one workface, it still maintained three productive work-faces; and

- it reduced the predictive contract period significantly.

Taking the facts as they were, the model showed that even with all the delaying factors that had actually occurred, including the failure to make pipe deliveries to the river-head camp, the introduction of a third base camp and its two additional work-faces would have cost no more than could have been predicted to be likely to be incurred as a result of foreseeable prolongation costs which could have been predicted at tender stage. But, perhaps more significantly, it would have reduced the ultimate site work period by 18 months, a reduction in time which could have resulted in many millions of dollars in revenue being received earlier than they were, with a proportionate reduction of project financing costs.

Lessons learned

In a contract that requires the employer to compensate the contractor with more time and money for delay caused by, or at the risk of, the employer, it is unhelpful to leave the strategic planning to a contractor who is motivated to bid the lowest contract time and cost, irrespective of the out-turn time and out-turn cost.

There is a significant disincentive to the tenderers to take into consideration the likely prolongation costs arising out of the employer's expected risks. A more productive approach would have been for the employer to require all tendering contractors to include the costs of three base camps positioned at approximately equal distances along the ROW.[2]

Case study 2: Railway in Australia

Project outline

There was an existing railway line to the north and south of a State-owned industrial park, linked by a railway track that crossed a four-lane highway at grade (a level crossing). When first constructed there was little vehicular traffic on the road and little rail traffic requiring the use of the junction at grade. Accordingly it functioned reasonably well by simply lowering the road barrier and stopping vehicular traffic when the rail link was needed. However, over time, the road had developed as a principal commuter route into the city and the railway had become busier, resulting in considerable congestion at peak travelling periods.

It was decided to build a new rail link through the State-owned industrial park with an elevated road section over the rail link. This would obviate the need for the crossing at grade (which could then be closed), relieving both the road and rail congestion.

[2] Right of Way.

The contract

The State's standard form engineering contracts required the employer to grant an extension of time and to pay compensation for prolongation or delay caused by anything outside the contractor's control. This applied equally to build-only and to design-and-build contracts. Typically, these include:

■ instructed variations;

■ implied variations arising out of wrongful conditional approval or wrongful rejection of submissions;

■ rebalancing of prime cost and provisional sums;

■ expenditure of contingency sums;

■ correction of errors in or omissions from bills of quantities;

■ remeasured approximate quantities;

■ unknown services or obstructions below ground;

■ delayed provision on information, instructions or approvals;

■ delayed provision of design information;

■ exceptional weather conditions;

■ suspension of work as a result of employer's default;

■ any default on the part of the employer or anyone under the employer's employment or control; and

■ force majeure, and so on.

The strategic plan

The strategic plan was to appoint a design-and-build contractor for construction of the flyover, who could bring to the project innovative construction techniques and keep the road traffic flowing during construction. The employer could then design its own signalling and new railway link to pass under the completed flyover. This was to be carried out by one or more build-only contractors. It was thought that by leaving the rail work until last, this would keep the rents from the industrial park flowing into the employer's coffers for as long as possible, thereby making the work less expensive than it would otherwise be.

The contractor's planning solution involved accessing the site of the flyover from the north via the industrial park, closing the south side of the highway and diverting the already congested traffic onto the remaining northern single lane in each direction. The south side of the highway could then be broken up and the south side of the flyover constructed in its place, complete with tie-ins to the existing highway. Having completed the south side, the traffic would then be diverted onto the new south flyover and the process repeated on the remaining two lanes of the north highway.

The challenge

The strategic plan presented the following challenges, which rendered it unlikely to be a successful solution:

■ The single critical path of the proposal (north flyover, north road tie-ins, south road flyover, south road tie-ins, rail track, north rail tie-ins, south rail tie-ins, signalling) meant that if there was any delay in completion of any phase, there would inevitably be a delay in commencement of the next, and so on, with the result of a delay to project completion.

■ Building the flyover while keeping the road open inevitably resulted in an already badly congested road becoming even more congested for a long period.

■ Because access was only to be made available from the north via the existing industrial park, the first phase of the flyover had to be constructed by working over a congested highway.

■ Large-scale protective screening would have to be constructed around the congested highway in order to protect the traffic. This would have to be carried out when the road could be closed, overnight in short possessions. It would also have to be maintained and repaired during short possessions, overnight.

■ Tie-ins are always problematical. They always require the existing congested routes to be closed during their construction, which usually means that the work has to be carried out at night in short possessions.

■ The proximity of the flyover and rail-link to the industrial park raised the likelihood of unknown services, or known services in unknown positions, hampering construction.

■ High, heavy-duty cranes necessary for the flyover construction would be unusable in high winds.

There was thus a high likelihood of extensive delay to completion and high likelihood of a claim for high prolongation costs. Conservatively, these were estimated to be likely to be about 12 months delay and $250m compensation, respectively.

A more effective planning solution

The key to relieving the pressure on a single critical path was to let an enabling contract to construct a temporary four-lane highway through the industrial park. This would inevitably put a stop to the income derived from the rents and create a new expenditure budget in the construction of the temporary road. However, the temporary road did not have to be of high quality, vehicular speed on it could be restricted, and it only had to last for a year or two at most.

The diversion of vehicular traffic onto the temporary road would mean that the contractor could then use the existing highway for access to both sides of the flyover site, obviating the need for special highway protection and high heavy-duty cranes, and rendering it possible to construct the whole flyover in one operation: a much simpler, quicker and cheaper process.

Further improvements in strategic planning could be obtained by constructing all the tie-ins (both rail and road) in parallel, while the flyover was being constructed, allowing a quick completion to the full project scope shortly after the flyover was finished.

Conservatively, this revised strategy was expected to produce a three months contingency period for any delay in the construction of the tie-ins (thereby rendering the potentially most problematic part, least likely to cause delay) and to reduce the overall planned construction period by 12 months, with an associated reduction in contract site costs and overheads to offset the loss of rent revenue and the cost of the enabling work.

Ultimately, the advantages of the more effective planning strategy were:

■ the absence of increased traffic congestion during construction;

■ shorter construction period and quicker relief of traffic congestion;

Appendices

- simpler construction processes, resulting in less likelihood of construction delays;

- parallel paths with less risk of delay to project completion;

- significant contingency periods for delay could be included on the most problematic sequences; and

- more certainty of completion on time and less likelihood of prolongation cost claims.

Lessons learned

In a contract that requires the employer to compensate the contractor with more time and money for delay caused by anything conceived to be outside the contractor's control, it is unhelpful to have a single critical path, from start to finish.

It is also unhelpful to plan the work without any contingency periods for delay at the employer's risk.

Case study 3: Administrative offices in Europe

Project outline

An existing office block of several hundred thousand square feet was to be extensively refurbished and extended. A preliminary study had ruled out the possibility of finding temporary accommodation and vacating the entire existing building while the work was carried out.

Contract

The employer's standard form contract was adapted with project-specific, special conditions which required the employer to grant an extension of time and to pay compensation for prolongation for delay caused by those matters identified as the employer's risks. Typically, these include:

- instructed variations;

- implied variations arising out of conditional approval or wrongful rejection of submissions;

- rebalancing of prime cost and provisional sums;

- expenditure of contingency sums;

- correcting errors in or omissions from bills of quantities;

- remeasured approximate quantities;

- unknown services or obstructions below ground or in existing structures;

- delayed provision on information, instructions or approvals;

- delayed provision of design information;

- unusual weather;

- suspension of work as a result of employer default;

- any default on the part of the employer or anyone under the employer's employment or control; and

- force majeure, and so on.

The strategic plan

It was determined that once the extension was complete, the refurbishment could be carried out in four, roughly equal phases and that the extension could accommodate the decanted personnel from each refurbishment phase, sequentially.

The planning strategy thus involved designing the extension for a build-only contractor to carry out the works. On completion of the extension design, while the extension was being built, the design of the first phase of refurbishment could be completed so that, on completion of the new extension, the personnel in the first phase of the refurbishment could be decanted into the extension in preparation for the first phase refurbishment contract to be executed on site.

Whilst that first phase refurbishment contract was on site, the second phase refurbishment could be designed so that, on completion of first phase refurbishment, personnel could be decanted from the extension back into the refurbished part and personnel from the second phase refurbishment could then be decanted into the new extension, in preparation for the start of the second phase refurbishment on site. The project could then proceed similarly through the third and fourth phases, to completion.

The whole was expected to take ten years to complete

The challenge

In effect, the strategy involved a single critical path commencing with the design of the extension, through the completion of construction of the extension, followed sequentially by the completion of four, roughly equal-sized refurbishment programmes.

If the design of the extension was late in completion, the construction of the extension would be late starting, and hence late in completion. The same would apply to every other phase. Although there was some float in the design of each refurbishment programme, refurbishment is notoriously difficult to control in both design and construction simply because much of the work that will ultimately be needed is likely to be in concealed areas which cannot be fully designed until it has been opened up and inspected.

Accordingly, the chances of the scheme (as designed) being completed on time and on cost were negligible. Conservatively, the project was thought to be likely to finish about 18 months late. Although there was no heavy engineering involved, prolongation costs for five contractors could reasonably be expected to result in claims in the region of €200m.

A more effective planning solution

The design and construction of a large office block in close proximity to other office buildings should not present any difficulties, which are not reasonably foreseeable. That is not to say that nothing will happen to delay completion, it just means that some delay to completion is reasonably foreseeable as a result of the occurrence of all those matters at the employer's risk under the contract.

Refurbishment projects are notoriously difficult to predict and not only bear the same risks as new construction on a green-field site, but also bear a large number of 'refurbishment specific' risks and an enhanced likelihood of them occurring to delay the project.

To make the timely project completion more certain and reduce the risk of substantial prolongation costs, a method must be found to create float in a path that is likely to be delayed. This could be achieved by making the first refurbishment phase not reliant on the completion of the extension, either by building temporary accommodation into which the first refurbishment phase personnel could be decanted during the construction period (maybe an inflatable building) or finding suitable temporary accommodation to rent, of a size sufficient for a single refurbishment phase.

The effect of this would be to enable the first phase refurbishment design to proceed in parallel with the design of the new extension so that the first phase refurbishment on site could commence while the extension was still being built, if necessary. On completion of either the first phase refurbishment or the extension (whichever occurred first), the second phase refurbishment personnel could be decanted into it, and so on. The effect of this would be to shorten the predicted overall construction period by about three years with a concomitant reduction in the contractor's time-related costs and overheads that would more than cover the minimal costs of temporary office accommodation.

Lessons learned

It is normally the case that a single critical path from commencement to completion will result in delayed completion. It is less risky and less costly to find a way of achieving more than one work sequence, which will facilitate parallel processing and, in the event of delayed progress on the critical path, allow a shift to a different sequence that will permit work to continue.

Case study 4: A viaduct in Asia

Project outline

A particular section of a road improvement scheme required a high-level, four-lane viaduct to be constructed over an existing village. The scheme involved the construction of concrete pylons from which were suspended concrete post-tensioned concrete box-sections to form the four-lane viaduct.

At the eastern end of the planned viaduct, the viaduct was crossed by high-voltage cables that were too low to permit the construction gantry to travel beneath them while the cables were live. It was established that power could only be shut down for about 6 weeks in total between February and March in any one year.

Independent consulting engineers prepared the concept design for the work under a contract that required the contractor to carry out the detailed design to the engineer's approval under a build-only contract.

The contract

Construction was carried out under the State's standard form of engineering contract, which required the employer to grant an extension of time and to pay compensation for prolongation caused by matters identified as employer risks. Typically, these include:

■ instructed variations;

■ implied variations arising out of wrongful conditional approval or wrongful rejection of submissions;

■ rebalancing of prime cost and provisional sums;

■ expenditure of contingency sums;

■ correction of errors in or omissions from bills of quantities;

■ remeasured approximate quantities;

■ unknown services or obstructions below ground;

■ delayed provision on information, instructions or approvals;

■ delayed provision of design information;

■ typhoon;

■ suspension of work as a result of employer's default;

■ any default on the part of the employer or anyone under the employer's employment or control; and

■ force majeure, and so on.

The strategic plan

The contractor planned to have the concrete box sections for the road elements pre-cast off-site and then ferried to the site on an as-needed basis, the first cast being the first needed for erection, and so on. The pylons were independent structures, each of which could be constructed at any time in any order. A single travelling gantry was to be employed to hoist the box sections into position, starting at the west end and proceeding to the east. Before the gantry could hoist any boxes into position, the pylons before and after the pylons between which the box sections were to be erected had to be completed.

The challenge

If erection of the concrete box sections by the gantry was delayed, and the gantry could then not reach the power lines section at the east end by February, that section could not be completed until the following year, and would cause delayed completion.

A number of difficulties caused the gantry's progress to be delayed, some at the employer's and some at the contractor's risk. In order to avoid the problems of limitation on the period of the power-line outage, the contractor obtained many more lifting rigs to enable work to continue on other pylons whilst the gantry was held up. Eventually, the gantry was able to recommence and proceeded to the east end about a year later than planned. In total the road was completed about 18 months late, with claims for additional cost amounting to twice the original contract value.

A more effective planning solution

The single critical path for concrete box section casting, delivery, and erection from west to east, was always going to be problematic, simply because of the narrow window of opportunity to pass under the power lines at the far end of the viaduct, and presented, by the west end start, the maximum opportunity for things to go wrong. It would have been potentially less problematic if gantry erection had been planned to start at the east end, at which there would have been significantly less likelihood of severe interruption to the process before encountering the power lines.

The linear nature of the construction process above the pylons meant that if anything delayed the construction of the pylons in the correct order, or prevented the casting, delivery or erection of the concrete box sections in the correct order, the whole of the work would be held up.

In order to minimise the risk of delayed completion, it would have been necessary to employ two erection gantries, one starting at the west end and proceeding east, whilst the other starting at the east end and proceeding west. In this way if either gantry was delayed in erecting any of the sections, the other could carry on past the mid-point, thereby reducing or eliminating potential delay to completion.

The use of two gantries instead of one would also shorten the construction period significantly and reduce the contractor's site costs and other time-related overheads.

Lessons learned

Whenever there is a potential restriction on continuous progress, as in this case, it makes sense to carry out and complete the work through the restriction as early as possible in the programme of work because there will then be less opportunity for interference to cause a serious delay to completion.

The single critical path of box-section erection could have been avoided by the use of two gantries, one working from west to east and the other from east to west. In this way a delay to one could be expected to be mitigated to some extent by the progress of the other, and would not necessarily result in delayed completion. The cost of the second gantry would be offset in part by the reduced contract period arising out of the doubled productivity and consequential reduction in site preliminaries and overheads.

Summary

In none of the case studies had any time contingencies been allowed in the strategic plan for the potential delaying effect of project-specific risks. Although cost contingencies were included in the budgets, no time to spend those contingencies was identified in the planning strategy and no allowance had been made in the budgets for prolongation costs.

In other words, although the contract time and cost were ascertained, the out-turn time and cost assumed no risk events would occur to delay progress, a highly improbable scenario.

Had priced time contingencies been included in the strategic plan, it would have been obvious that the planned contract period was unachievable within the boundaries of the foreseeable risks and the planned strategy, and as a consequence the project would be overspent. In other words, as planned, all these projects were doomed from the start.

The possible objections to planning to absorb delays to progress without delaying completion by the use of multiple parallel paths and strategic use of contingency time are:

■ *Parallel paths increase the rate of spend.*

This is true and inevitable.

■ *Working in parallel creates too much work for the existing management team.*

The answer to that is to double up on the management teams.

■ *Doubling the management team would cost too much.*

That is highly unlikely. At first, it might be thought that by carrying out two operations at once it would be necessary to double the management costs and hence lose any possible savings. Except in the rare circumstances that consultancy fees and management costs become cheaper in time, it is manifestly more cost-effective to have double the management costs over half the contract period.

■ *Showing that the work is likely to be delayed and allowing time and cost for it only increases the likelihood of claims.*

This assumption is a mistake. The effect of pre-budgeting for claims if they arise has the effect of focusing attention on risk and encourages a reduction in the risks experienced rather than encouraging them. Planning on the assumption that nothing will change, no information or drawings will be late, no submission will be wrongly rejected, no adverse weather or ground conditions will be found, nothing will break down or be delivered late, and all labour will turn up on time and work efficiently, etc., is a theory that has no historical foundation nor relevance to major projects. The larger and more complex the project and the longer the programme of development, the more all types of delaying events are likely to occur. Only by making due allowances and planning contingencies for them can their consequences be managed properly and the best way of achieving this is by means of effective strategic planning against a dynamic time model.

■ *None of the standard forms of contract permit the adjustment of time contingencies.*

That is incorrect. The Irish Government standard forms of construction and engineering contracts and the CIOB Time and Cost Management Contract (2015)[3] both provide for time contingencies and their adjustment under the contract, although they deal with them in different ways.

■ *None of the standard forms of contract require a dynamic time model; they all call for a static baseline target programme.*

That is incorrect, the CIOB Time and Cost Management Contract 2015[4] requires a variable baseline target in a dynamic time model and does not permit static baseline target programming.

Appendices

---

[3] http://www.ciob.org/insight/time-and-cost-management-contract-suite (accessed 10 December 2017).
[4] Ibid.

APPENDIX 3   The nature of complex projects

Complex projects can be defined both inclusively and exclusively. Much will depend upon the perceptions of those involved as to whether, for the purpose of time management, a particular project is complex or not. On the other hand, experience and the results of the CIOB's research indicate that the following classifications are likely to prove helpful.

Simple projects

Projects which are relatively simple from a time management perspective comprise those in which construction has all the following characteristics:

- design work is completed before construction starts;

- work comprises a single building (or repetition of identical buildings);

- construction is lower than 5-storey height;

- without below-ground accommodation;

- carried out to a single completion date;

- without phased possessions or access dates;

- with services not exceeding single-voltage power, lighting, telephone, hot and cold water, and heating;

- with a construction period shorter than nine months;

- with a single contractor; and

- with fewer than 10 subcontracts.

Complex projects

From a time-management perspective, complex projects tend to comprise those in which construction has any one or more of the following characteristics:

- design work is to be completed during construction;

- work comprises more than one building;

- construction is higher than 5-storey height;

- contains below-ground accommodation;

- to be completed by multiple key dates and/or sectional completion dates;

- with multiple possessions or access dates;

- with short possessions;

- work contains services exceeding single-voltage power, lighting, telephone, hot and cold water, and heating;

Guide to Good Practice in the Management of Time in Major Projects: Dynamic Time Modelling,
Second Edition. The Chartered Institute of Building.
© 2018 John Wiley & Sons Ltd. Published 2018 by John Wiley & Sons Ltd.

- construction work is accompanied by work of civil engineering character;
- the construction period is longer than 12 months;
- construction is to be carried out by multiple contractors or trade contractors;
- by more than 20 subcontractors.

APPENDIX 4   The dynamic time model – a flow chart

Appendices

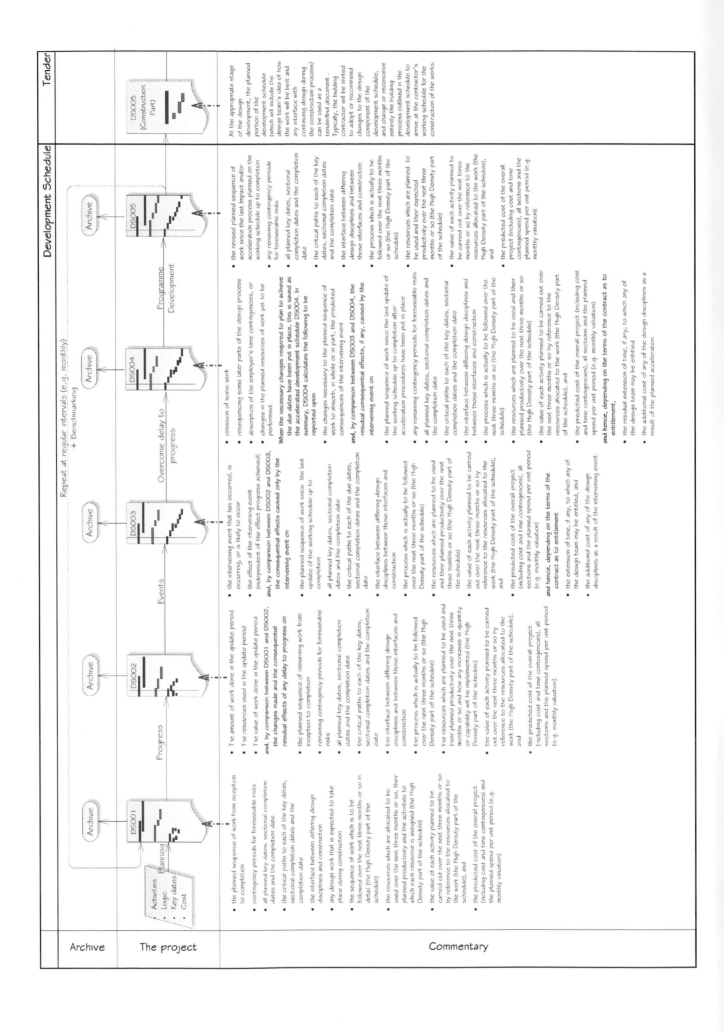

Tender | **Development Schedule**

Repeat at regular intervals (e.g. monthly) + Benchmarking

Flow: Planning → Progress → Events → Overcome delay to progress → Programme Development

Boxes: DS001 · DS002 · DS003 · DS004 · DS005 · DS005 (Construction Part)

Planning inputs: Activities · Logic · Key dates · Cost

Left labels: **Archive** / **The project** / **Commentary**

DS001

- the planned sequence of work from inception to completion
- contingency periods for foreseeable risks
- all planned key dates, sectional completion dates and the completion date
- the critical paths to each of the key dates, sectional completion dates and the completion date
- the interface between differing design disciplines and construction
- any design work that is expected to take place during construction
- the resources which are allocated to be used over the next three months or so, their planned productivity and the activities to which each resource is assigned (the High Density part of the schedule)
- the value of each activity planned to be carried out over the next three months or so by reference to the resources allocated to the work (the High Density part of the schedule), and
- the predicted cost of the overall project (including cost and time contingencies) and the planned spend per unit period (e.g. monthly valuation)

DS002

- The amount of work done in the update period
- The resources used in the update period
- The value of work done in the update period and, by comparison between DS001 and DS002, **the changes made and the consequential residual effects of any delay to progress on**
- the planned sequence of remaining work from inception to completion
- remaining contingency periods for foreseeable risks
- all planned key dates, sectional completion dates and the completion date
- the critical paths to each of the key dates, sectional completion dates and the completion date
- the interface between differing design disciplines and between those interfaces and construction
- the process which is actually to be followed over the next three months or so (the High Density part of the schedule)
- the resources which are planned to be used and their planned productivity over the next three months or so and how any increases in quantity or capability will be implemented (the High Density part of the schedule)
- the value of each activity planned to be carried out over the next three months or so by reference to the resources allocated to the work (the High Density part of the schedule), and
- the predicted cost of the overall project (including cost and time contingencies), all sections and the planned spend per unit period (e.g. monthly valuation).

DS003

- the intervening event that has occurred, is occurring, or is likely to occur
- the effect of the intervening event (independent of the effect progress achieved) and, by comparison between DS002 and DS003, **the consequential effects caused only by the intervening event on**
- the planned sequence of work since the last update of the working schedule up to completion
- all planned key dates, sectional completion dates and the completion date
- the critical paths to each of the key dates, sectional completion dates and the completion date
- the interface between differing design disciplines between those interfaces and construction
- the process which is actually to be followed over the next three months or so (the High Density part of the schedule)
- the resources which are planned to be used and their planned productivity over the next three months or so (the High Density part of the schedule)
- the value of each activity planned to be carried out over the next three months or so by reference to the resources allocated to the work (the High Density part of the schedule), and
- the predicted cost of the overall project (including cost and time contingencies), all sections and the planned spend per unit period (e.g. monthly valuation)

and hence, depending on the terms of the contract as to entitlement,
- the extension of time, if any, to which any of the design team may be entitled, and
- the additional cost as a result of the intervening event.

DS004

- omission of some work
- resequencing some later parts of the design process
- absorption of the employer's time contingencies, or
- changes in the planned resources of work yet to be performed.

When the necessary changes required to plan to achieve the due dates have been put in place, this is saved as the accelerated development schedule DS004. In summary, DS004 calculates the following to be reported upon
- the changes necessary to the planned sequence of work to absorb, in whole or in part, the consequences of the intervening event

and, by comparison between DS003 and DS004, the residual consequential effects, if any, caused by the intervening event on
- the planned sequence of work since the last update of the working schedule up to completion after acceleration procedures have been put in place
- any remaining contingency periods for foreseeable risks
- all planned key dates, sectional completion dates and the completion date
- the critical paths to each of the key dates, sectional completion dates and the completion date
- the interface between differing design disciplines and between those interfaces and construction
- the process which is actually to be followed over the next three months or so (the High Density part of the schedule)
- the resources which are planned to be used and their planned productivity over the next three months or so (the High Density part of the schedule)
- the value of each activity planned to be carried out over the next three months or so by reference to the resources allocated to the work (the High Density part of the schedule),
- the predicted cost of the overall project (including cost and time contingencies), all sections and the planned spend per unit period (e.g. monthly valuation)

and hence, depending on the terms of the contract as to entitlement,
- the residual extension of time, if any, the design team may be entitled
- the additional cost of any of the design disciplines as a result of the planned acceleration.

DS005

- the revised planned sequence of work since the last Impact and/or acceleration process planned on the working schedule up to completion
- any remaining contingency periods for foreseeable risks
- all planned key dates, sectional completion dates and the completion date
- the critical paths to each of the key dates, sectional completion dates and the completion date
- the interface between differing design disciplines and between those interfaces and construction
- the process which is actually to be followed over the next three months or so (the High Density part of the schedule)
- the resources which are planned to be used and their expected productivity over the next three months or so (the High Density part of the schedule)
- the value of each activity planned to be carried out over the next three months or so by reference to the resources allocated to the work (the High Density part of the schedule), and
- the predicted cost of the overall project (including cost and time contingencies), all sections and the planned spend per unit period (e.g. monthly valuation).

DS005 (Construction Part)

At the appropriate stage of the design development, the planned portion of the development schedule (which will include the design team's idea of how the work will be built, and any interface with continuing design during the construction process) can be used as a tendered document. Typically, the bidding contractor will be invited to adopt or recommend changes to the design component of the development schedule, and change or reconceive entirely the building process outlined in the development schedule to arrive at the contractor's working schedule for the construction of the works.

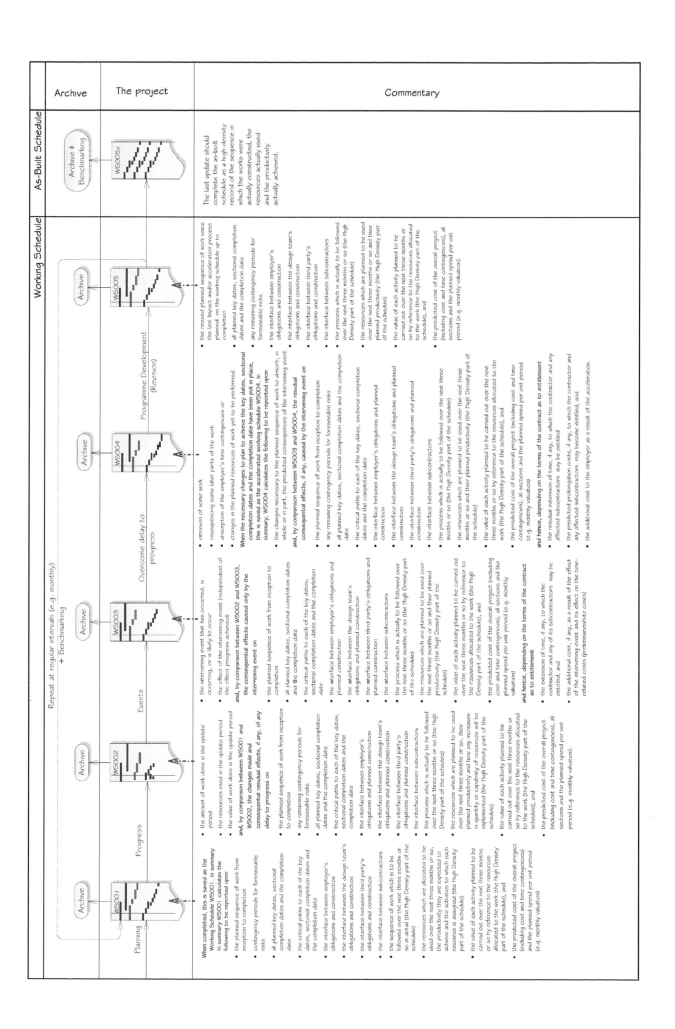

APPENDIX 5   Case studies in high density scheduling contents

Introduction

This appendix illustrates the time, control and cost advantages of an integrated, resource-based, High Density schedule. A static baseline, target programme, prepared for the whole project at the bid stage is the antithesis of this Guide. However, an integrated, resource-based, High Density schedule is essential for the optimisation of resource usage and the close control of the risk of interference to the work to be carried out in the near future. The case studies are:

1. Piling within an existing railway station.
2. Preparation of a piling schedule at tender/bidding stage.

Necessarily, these case studies have been selected from experience of many projects over many years. None describe the circumstances of any one project, and in each some facts have been changed to make the illustration simpler and easier to understand. Accordingly, any similarity between these examples and any real project is unintentional.

Case study 1: Piling within an existing railway station

Project outline

This case study concerns the scheduling of the piling element of a construction contract for the extension and refurbishment of a railway station while minimising disruption to the continuing rail services.

The piling for this project was particularly complicated in that it had to be carried out inside the existing vaulted structures and access was only available in limited possessions.

The piling subcontractor was constrained by space and time, which in turn had an effect on:

- the type of plant and equipment that could be used (restricted height and location of work);

- the sequence (handover to work areas as the pre-piling works in each location were completed); and

- the obligation to keep the station fully operational.

The contractor's schedule

The subcontractor was issued with both Low Density and Medium Density piling schedules prepared by the contractor. However, there was no High Density dynamic schedule in place to keep track of progress achieved in relation to what was planned, or to calculate the effect of intervening events on the progress of the work. The piling work had commenced on site using a short-term look-ahead plan prepared on a spreadsheet, and had soon fallen into delay.

Guide to Good Practice in the Management of Time in Major Projects: Dynamic Time Modelling,
Second Edition. The Chartered Institute of Building.
© 2018 John Wiley & Sons Ltd. Published 2018 by John Wiley & Sons Ltd.

Appendices

Because of the vagueness of the spreadsheet, it was not possible to see why the piling subcontract was in delay, and it soon proved to be insufficient for the management of the complex piling operations.

It was striking that the piling contractor employed no full-time scheduler and relied mainly upon the experience of its staff and simple spreadsheets for planning, monitoring and reporting projects.

The piling subcontractor realised that it needed expertise beyond those it had used in the past and employed a consultant scheduler in the first instance, to create the tools and plan for this project.

The subcontractor's schedule

The subcontractor was required to produce the High Density, fully resourced critical path piling schedule for integration with the contractor's schedule.

The subcontractor's High Density schedule was to be calculated from the optimum output of the various resources and to be prepared as a dynamic critical path network schedule that would respond to the changes that were likely to be required in the future as the result of instructed change and other events.

The challenge was to design a schedule that would provide the piling team and contractor with a tool that would both monitor progress and forecast likely outcomes whilst optimising resource usage (i.e. calculating the optimal utilisation of the specialised, restricted-access, piling-rigs).

The scheduling solution

Key to this solution was interpretation of the main contractor's Medium Density schedule, which defined the general sequence and geographical direction of the piling areas – piling direction east to west across the site with the critical handover dates to certain pile cap areas taken into account.

From a scheduler's point of view this required a classic, resource-driven schedule in which the resource logic is derived from the optimal number of rigs (and crews) available to deliver the sequence of piles planned in the relevant locations.

The schedule needed to illustrate the piles to be completed in order from east to west across the site in order to achieve completion of each section by the contractual handover dates, but logical links between activities would not be identified until after the resourcing schedule had been determined.

The following resource details were required from the subcontractor:

■ maximum number of rigs available;

■ optimum output per piling rig in number of piles driven per hour;[1]

■ any variations between individual rig outputs;

■ working time and any limitation on possession period;

■ number of working shifts per day;

■ number of working hours per shift.

The planning software permitted a 'pool' of the maximum number of available rigs to be created, from which the software optimally allocated the piling rigs to the piling

---

[1] In this case, as the work had already started, the productivity rate to be used was the productivity actually achieved on site. On a project that had not started, the productivity rate would have to be theoretical, until it could be corrected by factual data.

activities based on their priority and allocated them according to the timed sequence required by the contractor's schedule. The number of rigs in the 'pool' were sequentially reduced until the minimum number of piling rigs could be allocated to deliver the required piles according to the schedule, producing a resource-optimized schedule with no logic links between activities.

Following agreement of the resource optimization, logic links were added so that the impact of any later changes on the schedule could be calculated.

As the work progressed, the High Density, fully-resourced, critical path network was updated daily with actual productivity and sequences. This provided real, live progress data upon which management decisions could be made to optimize delivery performance.

Lessons learned

Complexity was a very important ingredient in this project, which had not been taken into account sufficiently in the contractor's initial piling schedule or the subcontractor's spreadsheet planning. The required priority of work sequence, together with severe site restrictions and handover dates, meant that simplistic planning would never be sufficient to deliver the required piling work efficiently and effectively by the completion dates required.

High Density scheduling should take place in good time before commencement of work on site. A more formalised, quality-controlled scheduling process would have helped tease out the time-management and resource requirements from the outset.

Case study 2: High Density network at tender/bidding stage

Project outline

The project was the construction of a large, complex, new multistorey building forming the anchor element of a major redevelopment. The subcontractor was required to price the piling element of the project for inclusion in the contractor's bid for the construction contract.

If successful in its bid, the piling would be one of the first activities to be carried out. The subcontractor had learnt the importance of good quality scheduling (see Case Study 1 above) and employed an experienced scheduler to provide the subcontractor's proposed piling schedule and to advise on the timing element of the piling in the final stages of this bid.

The contractor's schedule

The piling was not technically complex, other than the piles were designed to be large diameter and very long. The main contractor had provided a series of diagrams indicating the planned sequence of works and its interpretation of the likely piling schedule.

The key requirement here was for the piling contractor to establish the most cost-effective use of time and resources to enable a competitive bid to be calculated.

The subcontractor's schedule

The challenge was to effectively translate the piling contractor's technical expertise into a cost-effective schedule that would meet both the piling contractor's and the main contractor's financial goals.

The subcontractor's initial plan was not in a form that was a dynamic logic-linked schedule and was not acceptable to the main contractor.

To remedy this deficiency, the scheduler worked very closely with the subcontractor's technical and engineering staff based on the main contractor's strategic work sequence.

A High Density, resource-loaded schedule was feasible at this stage of the project because the number of piles, their design and positions were all known and the strategic sequence of installation was split into zones across the site. The activities were grouped into their zones and the logic determined by the main contractor's strategic building sequence drawings.

The key here was to determine the optimum number of piling rigs in each location and the optimum working sequence. The scheduler needed to know the following resource details:

- maximum number of rigs available;

- optimum output per piling rig in number of piles driven per hour;[2]

- any planned differences between individual rig outputs;

- planned working time and any limitation on possession period;

- planned number of working shifts per day;

- planned number of working hours per shift;

- identification of zones of work;

- the number and type of piles required for each zone.

The number of piles required was used as the resource output and the piling rigs were allocated accordingly, depending on pile type, location and rig availability. Their planned efficiency was identified by resource histograms illustrating under- or over-utilisation of the rigs. Through an iterative optimisation process (using deadline critical resource analysis) the minimum number of rigs that could be used to meet the contractor's scheduled dates was determined.

The final number of rigs calculated to be needed was actually one less than the piling subcontractor had previously calculated to be required using his previous spreadsheet planning method.

He remarked that, after five years' experience of planning with spreadsheets, the number of rigs needed was more accurately calculated in the last two weeks of the bid process by use of 'proper' scheduling knowledge and the appropriate tools.

Lessons learned

In this case the piling contractor had learned from Case Study 1 above the importance of good scheduling, and had recognised that it did not have sufficient in-house expertise to:

- produce schedules suitable for the bid process; or

- plan in such a manner to optimize both technical and financial performance.

The company now employs a full-time scheduler.

---

[2] Obviously, as the work had not started, the productivity rate to be used was theoretical and initially based upon the subcontractor's experience of other projects. Once started on site, the theoretical output could be corrected by factual data from the progress records.

Summary

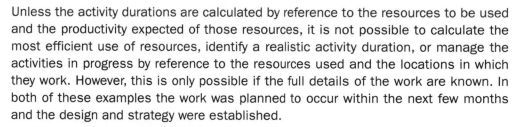

Unless the activity durations are calculated by reference to the resources to be used and the productivity expected of those resources, it is not possible to calculate the most efficient use of resources, identify a realistic activity duration, or manage the activities in progress by reference to the resources used and the locations in which they work. However, this is only possible if the full details of the work are known. In both of these examples the work was planned to occur within the next few months and the design and strategy were established.

An independent, short-term look-ahead programme that is not integrated with the working schedule is of limited use in identifying the effect on those resources of any interruption to progress.

Appendices

Summary

APPENDIX 6   Desirable attributes of scheduling software

Primary considerations

Projects and subprojects

Software which can only cope with a single project at a time is unlikely to be sufficiently flexible for major projects, apart from the possibility of identifying separate sections as subprojects (which are subject to sectional completion, or separate key dates), or the work of separate contractors or subcontractors. For ease of application in practice, it may also be useful to identify separate operational zones as subprojects.

Whenever the schedule is constructed as a master schedule with synchronized subprojects, it is important that the subproject[1] should be accessible by other named parties via the internet or by access to a CDE in whole or in part, for viewing only, or for editing with security-dependent access rights.

The software should provide for the unique identification of each project file and include version control capabilities. The ability to compare versions of the same project file to identify the effect of changes is desirable.

The software should facilitate the logical linking of separate project files to allow the creation of a dynamically linked master schedule for the full project, whilst permitting the synchronised management of individual subprojects.

Activities

For each activity there should be:

- a unique activity-identifying alpha-numeric ID code;[2] and
- a unique activity description.[3]

Software that permits duplication of activity IDs or activity descriptions, without warning, is likely to produce schedules which lack clarity and are thus incompatible with good practice. The software should not facilitate that duplication at all, or, if it does, have a clear, permanent warning on the schedule as to the deficiency.

The software should be capable of distinguishing between the following activity and event types:

- duration-identified activities
- resource-calculated activities
- hammocks

---

[1] See 'Schedule integration' at Section 4.8.
[2] See 'Activity Identifier coding 'at Section 4.15.
[3] See 'Activity description' at Section 4.16.

Guide to Good Practice in the Management of Time in Major Projects: Dynamic Time Modelling,
Second Edition. The Chartered Institute of Building.
© 2018 John Wiley & Sons Ltd. Published 2018 by John Wiley & Sons Ltd.

179

Appendices

- start milestones or flags

- finish milestones or flags

- employer-owned contingency/risk allocation

- contractor-owned contingency/risk allocation

An activity-related field capable of taking free text and numbers as comments or notes is often a useful facility.

The software should be capable of identifying activity durations in different formats. Although for most purposes in construction, activity durations in days may be sufficient, for the purposes of limited possessions, durations in hours and minutes and, in the Low Density part of a schedule, durations in weeks and months are often necessary.

The software should make it clear as to what unit of time it uses to carry out its calculations (i.e. days, hours, minutes or seconds). The best software calculates to the minute.

The software ought to be capable of identifying which activities are logically determined to be of a shorter duration than the applied logic and whether they are to be 'stretched' or 'not continuous' as a result of the logic.

Logical relationships

The software should permit a logical flow of work and prohibit the indication of relationships that are impossible to perform in practice. It should provide forward and backward passes and detect logical loops and open ends. Any software that fails to do this is likely to produce schedules that are incompatible with good practice. It should either not facilitate such defects at all, or declare a clear, permanent warning on the schedule as to the deficiency.

The software should be capable of identifying all variations of logical links, either individually or in combination. Software that limits the user to finish-to-start logic, or too few logical connections to any one activity, is unlikely to be useful. The software should permit at least two links from the start and finish of every activity.

The software should identify any inconsistency between logic and the activity durations to which the logic is applied.

Logic should be capable of being illustrated as 'driving' or 'non-driving' to any chosen point within the schedule.

Logic should distinguish between:

- engineering logic (the construction sequence with no resource constraints);

- resource logic (the construction sequence carried out with the available resources);

- preferential logic (the construction sequence with imposed constraints to modify the purely 'engineered' and/or 'resourced' construction sequence);

- logic linking zones and/or subprojects;

- additionally, the software should be capable of identifying fixed lag and the working calendar the lag is to adopt; and

- lag should be listed as a logical attribute.

Constraints

Manually applied constraints are likely to be useful on most projects. Those that are acceptable in a network, when correctly applied, are:

■ start-no-earlier-than a given date;

■ start-no-later-than a given date;

■ zero-free-float;

■ additionally, the software should be capable of clearly identifying when a manual constraint has been applied to an activity.

Some software facilitates the use of constraints which will manipulate criticality and inhibit the ability of the software to model time accurately. These are not acceptable in a schedule used to manage time. Those which, if available in a software product, should not be permitted to be used in a critical path network without a clear, permanent warning on the schedule as to their effect, are:

■ any combination of constraints that will fix the earliest and latest dates for any activity or milestone;

■ a mandatory start date;

■ a mandatory finish date;

■ zero total float.

Critical path

The software should be capable of identifying:

■ the longest path to completion;

■ the longest path to intermediate key dates or sectional completion dates;

■ logic and activities which are critical, separately from those which are not critical to one or more completion dates;

■ total float on each path;

■ free float on each activity, on each path.

The software should be capable of facilitating the tracing of a critical path or paths through the driving logic of each activity on the critical path to a particular completion date or key date, from time to time.

Calendars

The software should be capable of facilitating the use of a number of different working calendars for activities, resources and lags, each capable of identifying different:

■ working-week start day

■ working weeks and weekends

■ working days

■ working hours

■ holidays

■ standard calendars and exceptions

Appendices

Resources

The software should be capable of facilitating the use of a number of different resources to determine a realistic schedule based on the resources allocated to each activity.

The resource scheduling facilities may be implemented in numerous different ways and, if resource analyses are to be correctly interpreted, it is essential for the scheduler to be able to understand how any specific tool works. Accordingly, the software documentation should provide comprehensive information on the various algorithms and options available within the tool.

The software should be capable of identifying resource availability at project level by reference to:

■ resource ID or unique identifying code

■ name of resource

■ resource working calendar

■ resource availability levels by date or time period

■ normal and maximum overload levels for each resource

■ the cost rates for the resource to be defined in terms of:

    ○ normally available

    ○ unused

    ○ overloaded

    ○ overtime working

    ○ lump sum

The software should be capable of allocating resources to an activity to permit:

■ the allocation of several resources to each activity;

■ the start and end time for each resource to be nominated in relation to the activity duration;

■ additionally, the software should be capable of identifying a split activity duration caused by resource shortages;

■ the quantity of each resource allocated to the activity to be varied at defined times relative to the activity duration;

■ resources to be allocated as a quantity and distributed over the activity duration based on a nominated profile;

■ the option of either maintaining the activity duration or calculating the duration based on effort (within defined maximum and minimum resource levels).

The software should be capable of calculating resources by progressing from project start and with the facility for:

■ resource smoothing without end-date extension (i.e. by using only the available float to minimise resource overloads when all of the available float has been consumed);

■ resource smoothing with end-date extension (i.e. by using the available float to minimise resource overloads until all of the available float has been consumed and then, when the maximum overload threshold is reached, delay critical activities to keep resource demand at, or below, the maximum overload threshold level);

■ resource levelling without overloading (i.e. by levelling the resource demand to remain at, or below, the planned level by delaying activities and end date if necessary);

■ cost calculation (i.e. based on the resource cost parameters, calculate the cost of each activity and the aggregate cost for the project by time period and in total);

■ by progressing backwards from project finish, using the results of the project-start allocations, with the facility for restraining the pacing resources to the normal or maximum overload level.

The software should be capable of resource reporting to demonstrate:

■ resource aggregation (e.g. by summarising each resource by time period based on the activities' early start and late finish dates);

■ resource histograms illustrating the effect of unlimited and planned resource quantities by time period; these should be capable of showing:

  O normally available

  O maximum threshold

  O used

  O unused

  O overloaded

■ for each activity:

  O the controlling resource when an activity has been delayed during the resource levelling procedure;

  O the calculated cost of the resources used on the activity;

■ project-cost reports identifying:

  O the cost per time unit of the resources allocated;

  O the cumulative project cost (e.g. project costs may include unused resource costs and various contingencies);

  O resource reports identifying the activities allocated to each specific resource;

  O aggregate reports identifying resource utilisation and cost reports aggregated by a time unit (e.g. month) or by a schedule component (e.g. a WBS element).

The software should be capable of permitting resource allocation to be updated to:

■ record actual resource usage and actual costs;

■ adjust planned resource usage on activities in progress and future work without impairing the use of archived previous editions of the schedule.

The software should be capable of identifying resource capabilities by reference to:

■ skill type;

■ alternative resources (i.e. if the nominated resource is overloaded, automatically use a nominated alternative);

■ resource-breakdown structures (with aggregate reporting similar to the WBS);

■ stacked histograms;

■ alternative resource-levelling and smoothing algorithms with clear descriptions of their functionality.

The software should also be able to:

■ store resource baseline archives separate from the current schedule;

■ pool resources across multiple projects;

■ manually set activity and project priorities (to influence resource allocations during resource scheduling);

■ pool resources within a project (see alternative resources);

■ perform earned-value aggregations.

WBS and activity content coding

The software should be capable of identifying a work breakdown structure. Whilst a structure of eight levels should be the ideal, a structure of fewer than five levels is unlikely to be practical on a major project.

The facility for a broad variety of bespoke database fields that can be displayed is usually an essential requirement of schedules for major projects.

Organisation

The software should be capable of organising the layout in any combination of fields and attributes, sorting activity, logic, attributes and values in any field.

Filtering

The software should be capable of filtering the content of any layout by selection of the value of any field, or attribute (either alone, or in combination with other fields, or attributes) on the basis of:

■ equal to

■ containing

■ not equal to

■ not containing

and, where the fields contain values, in relation to those values there should also be the facility for selection:

■ between

■ not between

Filtering should also facilitate the use of Boolean 'AND' and 'OR' logical combinations.

Layout

The minimum available layouts should comprise:

■ bar chart without logic

■ bar chart with logic

- network diagram (ADM or PDM)

- line-of-balance diagram

- resource profile

- cost profile

The software should have the facility for creating and saving a variety of different combinations of fields and attributes, organised and filtered, as layouts for reporting purposes.

The timescale to which the layout is restricted to view should be identifiable to any duration and density during the period between 6 months prior to inception of the earliest project and 12 years after planned completion of the latest project.

Every layout should be printable as both hard copy and PDF.

As-built data

The software should be capable of identifying the factual data for each activity and resource as:

- actual duration

- start date

- finish date

- percentage complete

- remaining duration

- calculated cost

- actual cost

- certified value

- resources expended

- productivity quotient achieved to date

Updating with progress

The software must be capable of identifying a data date (by a straight line through the activity bars) with as built data to the left and as planned data to the right of the data date line.

The software must be capable of comparing progress against the currently agreed baseline, such that any delays and/or changes in activity sequencing are clearly demonstrated.

The software must be capable of recalculating the critical path or paths and the predicted early and late start and finish dates of all activities and resources against the data date, with the effect that:

- all activities indicated to have started or finished are indicated to have started, or finished, earlier than the data date;

- no activity is identified to have started, or finished, later than the data date;

- activities which are in progress at the data date are indicated to be due to finish on a date after the data date proportionate to their degree of progress in relation to their planned duration at the data date.

Inputting and editing data

The software should be capable of holding input data and edits in memory so that they are subject to an instruction to 'undo' and only saved on a positive instruction to do so.

Archiving

Files should be capable of being saved in compressed-data format for archival purposes.

Training and support

The availability of effective, product-related training is extremely useful even for experienced schedulers. Even with the simplest of software, it is always helpful to understand how the software supplier identifies that it should be used.

Because of the sophistication of modern software and the inability or unwillingness of the manufacturers to subject products to rigorous testing before release, committed and easily available software support and continuously updated software is more important today than it has ever been.

Secondary considerations

Those matters which do not add to the quality of the calculated output but to the manner of use and which, depending upon circumstances, may be of some importance, include the following.

Enterprise-wide software

Enterprise-wide software can directly link a project or projects across the internet such that large, major projects are able to be scheduled and effectively monitored across the world.

Enterprise-wide software can relate together all projects with which the company is concerned. It is a useful attribute for enhancing company management.

Communications

Whether the schedule can be accessed by other parties via the internet, in whole or in part, for viewing only, or for editing with security-dependent access rights, can be of importance in managing the schedule.

Appearance

Software capable of being customised according to company requirements for house style by using different fonts, line thicknesses, or type and colours for each available field, value in the database and/or the background is useful.

A drawing facility that can be used to highlight aspects of a report is often useful.

Comparison of schedules

For the purpose of identifying the effect of differences between schedules in the process of review, revision, updating and impacting causative events, it is useful to have the facility for comparing two or more schedules on a line-by-line basis. In practice this usually means the facility for identifying one or more target schedules that can be viewed simultaneously with the current schedule.

Appendices

Organisation

A facility for organising the layout in order of logical predecessors and successor activities is useful.

Transparency with other software

The facility for importing from, and exporting to, other scheduling software may be available, but if it is, it should be capable of listing the differences which result from such import or export.

The software should be capable of importing and exporting from and to other databases such as MS Excel and MS Access for updating, analytical and quality-assurance purposes. The facility for attaching hyperlinks to activity IDs should be available to enable the linking of such documents as photographs and videos, flow charts, procedures and planning and construction method statements, and progress records.

Integration with time-keeping and cost-keeping systems can facilitate automatic updating from time, plant and material records, which, in relation to a fully resourced schedule, can produce an automated update facility.

Risk analysis

The facility for stepping through a potential shift in timing of activities to ascertain the consequent shift in the critical path is useful.

Monte Carlo analysis will give a profile of likelihood of success against given criteria in a Low Density schedule, which, if accurately predicted against data that remain unchanged, will predict likely outcome.

Archiving

A backup capable of being set to default periods or to be executed manually is a useful facility.

Appendices

APPENDIX 7   Industry productivity guides

Reference may be made to such data as Planning Planet's output production rates,[1] or other industry standards.

Builders' pricing books will give an indication of durations as well as price for a multitude of different activities, for example, listed alphabetically. Similar 'price books' are available in many other countries, and a number of research papers on the subject are available via the internet, together with electronic databased estimating systems.

Notwithstanding the reassurance provided by recognised publications, these standards should be viewed with caution and the user should be careful to compare situations which are inherently job-specific with standard work descriptions and make appropriate adjustment.

Some of the guides used in the UK, Australia and the US are:

United Kingdom

- Laxton's[2]
- Spence Geddes[3]
- Spon's[4]
- Wessex[5]

Australia

- Cordell's[6]
- Rawlinson's Construction Costs Australia[7]

United States of America

- RSMeans Data[8]

---

[1] http://www.planningplanet.com/wiki (accessed 27 August 2016).

[2] *Laxton's NRM Building Price Book 2016, Laxton's SMM Building Price Book 2016*, http://www.laxtons.co.uk/LaxtonsSMM2016.php (accessed 6 November 2016).

[3] Spence Geddes, *Estimating for Building & Civil Engineering Work*, 9th Edition, 1996.

[4] Aecom, *Spon's Architects' and Builders' Price Book 2017, Spon's External Works and Landscape Price Book 2017, Spon's Civil Engineering and Highway Works Price Book 2017, Spon's Mechanical and Electrical Services Price Book 2017*, http://www.pricebooks.co.uk (accessed 6 November 2016).

[5] Bliss Books, *BCIS Wessex Comprehensive Building Price Book 2017*, 34th Edition, http://blissbooks.co.uk/index.php?route=product/product&product_id=1180 (accessed 6 November 2016).

[6] Cordell's *The commercial and industrial building cost guide (Australia)*, http://www.cordellestimating.com.au/ (accessed 6 November 2016).

[7] Rawlinson's *2016 Construction Cost Guide*, https://www.rawlhouse.com/construction-cost-guide.php (accessed 6 November 2016).

[8] *RSMeans 2017 Building Construction Costs Book*, https://www.rsmeans.com (accessed 6 November 2016).

Guide to Good Practice in the Management of Time in Major Projects: Dynamic Time Modelling,
Second Edition. The Chartered Institute of Building.
© 2018 John Wiley & Sons Ltd. Published 2018 by John Wiley & Sons Ltd.

APPENDIX 8   Sample notice of delay

Event Description   Instructions to resolve a discrepancy between design information

Event No.   25

Summary of issues   The walkway design contained a dimensional error in that the dimensions provided did not correlate to the coordinates on the drawing. The walkway steel was ordered from the dimensions provided. The walkway was fabricated and delivered. When it was installed it was short by 0.5 m. The Subcontractor amended the design at the Contractor's instructions of 20 May, to insert a small panel in mid length to accommodate the error. The walkway structure had first been installed between 1st April and 14th May 2015, however, because the corrections to the dimensional error inhibited its use, completion of the MOF Jetty was delayed until 11th June 2015.

Documents   Letter SC> C 11th May 2015
Letter SC> C 12th May 2015
Letter C > SC 20th May 2015
RFI. 51 – 6th June 2015
Working schedule at 11 May 2015

Sequence

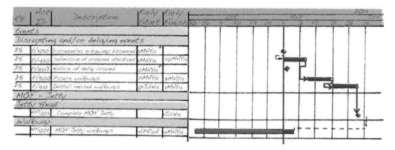

Contract   Clauses 44 and 51

Effect on time   Delay to Sectional Completion of MOF Jetty until 11th June 2015

Effect on cost   Prolongation of costs in late demobilization of sea-borne equipment and prolongation of MOF Jetty, details to be provided.

Guide to Good Practice in the Management of Time in Major Projects: Dynamic Time Modelling,
Second Edition. The Chartered Institute of Building.
© 2018 John Wiley & Sons Ltd. Published 2018 by John Wiley & Sons Ltd.

Glossary of terms

2D	Abbreviation for two-dimensional. A flat object, plan or model comprising of only length and width in plan, section and/or elevation
3D	Abbreviation for three-dimensional object or model comprising length, height and width
4D	The addition of the time schedule as the fourth dimension to a virtual model
5D	The addition of cost/value-related information as a fifth dimension to a virtual model
absenteeism	A failure of a labour resource to arrive at the designated workplace by the designated time
acceleration	Action to make up lost time or otherwise complete work at a rate quicker than that agreed in the approved schedule, usually at the employer's (q.v.) expense. See also 'recovery'
activity	A process consuming both time and resources
activity bar	A representation of the duration of an activity in a bar chart or linked bar chart in which a horizontal bar extends from the commencement date of the activity to the date of completion of the activity on the calendar scale
activity content code	A series of numeric and/or alphabetical characters used to identify the unique character of work associated with an activity, where each descriptive field (q.v.) has a specific code within the database. See Section 4.17
activity cost code	A series of numeric and/or alphabetical characters used to uniquely identify the value of work by reference to the cost of the required resources, where each descriptive cost-related field (q.v.) has a specific code within the database. See Section 4.18
activity description	A description of the work to be executed within an activity. See Section 4.16
activity duration	The number of time units required to complete an activity
activity ID	A series of numeric and/or alphabetical characters used to uniquely identify an activity. See Section 4.15
activity network	A graphical representation showing dependencies and other information in a schedule
activity scheduling	The process of scheduling by reference to activities planned to be carried out as opposed to the resources that are planned to be used and the location in which the resources will be employed. See also 'resource scheduling'

Guide to Good Practice in the Management of Time in Major Projects: Dynamic Time Modelling,
Second Edition. The Chartered Institute of Building.
© 2018 John Wiley & Sons Ltd. Published 2018 by John Wiley & Sons Ltd.

ADM	The abbreviation for 'arrow diagram method'
ALAP	The abbreviation for 'as-late-as-possible'. See Section 4.26.3.2
algorithm	A method of calculation
animation	A computerised virtual reconstruction of a sequence
approximate quantities	The estimated quantities of work, usually prepared from a scheme design, before detailed design is carried out, used for the purpose of a cost estimate that is subject to adjustment when the design is finalized and/or the work is carried out
area	A portion of the work, defined by reference to location or spatial coordinates, used for the purpose of management and control
archive	A storage respository of historic records, files and documents
arrow diagram method	The method of network scheduling in which the activity is represented by an arrow and events by nodes; see Section 4.10
ASAP	The abbreviation for 'as-soon-as-possible'. See Section 4.26.3.1
as-built	The work, as it has actually been carried out
as-built record	Drawings, sketches, measurements and progress data of the constructed/installed built asset. See Section 5.4
as-built schedule	The final schedule completed as a High Density (q.v.) record of the sequence in which the works were actually constructed, with the resources used and the productivity achieved. See Section 4.5.2.9
as-late-as-possible	See 'ALAP'
as-soon-as-possible	See 'ASAP'
audit trail	A sequence of checking whereby data in one document is validated before being relied upon in another. The facility for tracing the origin and sequence of changes made to a document, model or financial transaction
backfill	The material used or the process of using that material to fill an excavation
baseline	A fixed data point from which a departure can be established and measured
bar chart	An illustration of the project data with the activities represented graphically by bars set on a horizontal time calendar. Sometimes called a 'Gantt chart' (q.v.). See Section 4.10
benchmarking	The analysis of current performance data for the purpose of setting a standard for future performance. See Sections 3.14 and 6.6
bid	See 'tender'
bid schedule	See 'tender schedule'
bill of quantities	A document prepared for the purpose of pricing the work, usually according to defined rules that set out the measured quantity of work and describe the quality standard of materials and workmanship
BIM	The abbreviation for 'building information modelling' (q.v.)
BIM environment	The facility for multiple designers and technical specialists to work collaboratively throughout the project life-cycle, thereby improving communication and business decisions

buildability	The degree to which it is practical to manufacture and/or construct the works or a part of the works
building information modelling	The use of one or more virtual 3D (q.v.) models containing parametric data for the collaborative production of design information to describe the design of the works. See Sections 1.11, 2.11, 4.10 and 5.3
build-only contract	A form of project procurement in which the contractor is required only to build that which has been designed by the employer or those under the employer's control
buffer	See 'contingency'
CAD	The abbreviation for 'computer-aided design'
calendar	The time intervals during which activities may or may not be scheduled to be carried out, taking into account the available working days and non-working days, such as holidays and weekends. See Sections 3.7.15 and 3.8.17
cash flow	The expenditure, income, or balance of money received against money spent, plotted against time according to a defined formula
CBS	The abbreviation for 'cost breakdown system'
cell	A unit in a spreadsheet (q.v.) identified by horizontal and vertical positional alpha/numeric identifiers
change management	The art and science of controlling the effect of a departure from the contract quality, quantity, methodology, cost and timing of the work
civil commotion	Usually linked with 'riot' and referring to an uprising of a large number of people
clash detection	The process of identifying where the design models of the various specialists involved in the development of the project design have components or elements that occupy the same space, or other parameters that are incompatible, or, in 4D BIM modelling, a time sequence that is incorrect
coding structure	A method of categorising data by reference to values in fields (q.v.) of a database. See Section 4.15
CDE	The abbreviation for 'common data environment' (q.v.)
common data environment	A single source of information for collection, management and distribution of all relevant approved project documents for multi-disciplinary teams in a managed process compatible with the requirements of PAS 1192: Part 2, 2013 'Specification for information management for the capital/delivery phase of construction projects using building information modelling'
completion date	The end-date for the works, the subject of the planning and scheduling process
computer-aided design	The use of software in order to create precision drawings or technical illustrations in 2D (q.v.) or 3D (q.v.)
constraint	A restriction on the ability of a critical path network schedule to obey the activity and/or resource logic of the schedule. See Sections 4.27 and 4.30.7
construction management contract	A method of project procurement in which the management contractor is required to manage the works carried out by separate trade package contractors (q.v.)
construction manager	The title given to the person responsible for managing trade or subcontractor performance

construction method statement	A report, prepared by the contractor, identifying the means and methods intended to be used in constructing the works and temporary works, including any changes to it
contiguous duration	The gross period of time required to carry out an activity in an uninterrupted process
contingency	A provision of money, resource and/or time which may or may not be required; usually assessed to offset the consequences of identified risks, should they occur.
contingency activities	An activity representing allowance of time and/or resources to execute unforeseen work, interruptions to production or intervening events
contingency period	A period of time included within the development schedule or working schedule as an allowance for the time it is estimated that a risk will take up, if it occurs. See Section 4.9
contingency sum	A sum of money included within the contract price (or project budget) as an allowance for the cost it is estimated that a risk will take up, if it occurs. See Section 4.9.5
contract administrator	The title given to the party paid by the employer (q.v.) for administering the construction contract
contractual date	A date set out in the contract by which a specified event must be achieved, such as a sectional completion date or the completion date.
contractor	The title given to the contracting party responsible for constructing the works (q.v.)
contractor's contingency	An activity representing the contractor's allowance of time and/or resources to execute unforeseen work or interruptions to production at the contractor's risk
contract programme	The baseline programme (q.v.) agreed between the employer (q.v) and the contractor (q.v.) at the outset of the project, identifying the sequence and activity durations (q.v.) required to achieve the contract completion date.
consultants	Those employed by the employer (q.v.) in an advisory capacity; see also 'design team'
cost breakdown system	A method of categorising costs in order to aid efficient cost planning control during the project.
cost coding	See 'activity cost coding'
cost contingency	See 'contingency sum'
cost manager	The person or organisation identified in the contract with responsibility for managing the cost of the project within the authorised budget
count the squares	A primitive method of monitoring progress against a bar chart. See Section 5.9.24
CPM	The abbreviation for 'critical-path method'
crew	A defined group of labour (q.v.) usually tasked with a single trade
crew size	The quantity of labour in a defined group; see Section 4.20.13.6
critical activity	An activity that is on the critical path to one or more key dates, sectional completion dates or the completion date
critical path	The longest sequence of logically linked activities from the start of a schedule, or a part of a schedule, to its finish that determine the earliest possible achievement of a key date (q.v), sectional completion date (q.v) or the completion date (q.v.). See Section 4.29

critical path method	A technique used to predict project duration by analysing which sequence of activities has the least amount of scheduling flexibility. Early dates (q.v.) are figured by a forward pass using a specific start date, and late dates (q.v.) are figured by using a backward pass starting from a completion date.[1] See Section 4.10.2
critical path network	A schedule that identifies the planned sequence, interrelationships and dependancies between activities, the logic (q.v.) used, and calculates the critical path (q.v.). The schedule will normally be produced by a proprietary software product which can be configured to illustrate the schedule as an activity-on-arrow diagram (q.v.), precedence diagram (q.v), linked bar chart (q.v.), or bar chart (q.v)
critical path network analysis	A method of deducing the cause and effect of delays to progress by reference to the activities and their logical relationships in a critical path network (q.v.)
critical resource levelling	The distribution of currently available resources in order to allocate sufficient resources to permit critical activities (q.v.) to be carried out in the available time
CTS	The abbreviation for 'count the squares'
curing	The chemical process of hardening of a fluid mix of materials
cut and fill	A method of levelling land by 'cutting' from the high parts and using that product for 'filling' the low parts
D&B	The abbreviation for 'design-and-build'
dangle	See 'open end'
database	A structured set of data stored electronically in which data is recorded as values against fields (q.v.), which can then be searched, retrieved, reproduced and used in a multitude of forms by filter and organisation
database record	The data entered on a database. See Section 5.4.6
data date	The date upon which a schedule is revised or updated with progress data up to that date; also known as time now
data table	A listing of the data in a schedule by means of a horizontal list of each field (q.v.) and vertically, the value in each Activity ID. On the intersection is the cell containing the value per Activity ID of each field
deadline critical resource levelling	A method of strategic resource allocation in which potential resource overloading is reduced by moving activities within the available float (q.v.), if any, and if there is no available float, the allocation will show overloading of resources, but the project end date will not change
delay	In relation to a given work content, the difference between an intention and reality as to the timing of the work, where the reality is derogation from the intent
delay to progress	An adverse effect on the date by which the current schedule illustrates the start and/or finish of any identifiable activity, sequence, or resource
delivery lead time	The period between placing an order and receiving the goods
density design	See Section 4.2
density logic	See Section 4.24
dependency	The relationship between two or more activities which determines their respective timing and sequence of operation

---

[1] Wideman Comparative Glossary of Common Project Management Terms v.3.1.

design-and-build	A form of project procurement in which the contractor also carries design responsibilities
design brief	A written explanation of the employer's (q.v.) aims, objectives and key dates (q.v.) for the design of a project
design consultant	A member of the 'design team' (q.v.)
design programme	A programme (q.v) showing design development activities and possibly key dates (q.v) to be achieved. It may also show indicative construction activities and a completion date, but has not been created using critical path method techniques. Historically these have been found to be of limited value for managing risk and predicting future outcomes
design team	The title given to all those designing any part of the work but not including contractors who carry out design
development schedule	A critical path network (q.v.) schedule, prepared by or on behalf of the employer (q.v.) which identifies the planned process of project procurement from inception to completion, and which functions as the employer's time and cost model for the project as a whole
discretionary dependency	The logic (q.v.) of a part of a network (q.v.), which is based on a managerial preference rather than an engineering need, and which can be changed if necessary
dispute proceedings	The procedure of a formal tribunal to which a dispute has been referred for resolution, e.g. adjudication, litigation or arbitration
disruption	A disturbance or interruption to normal working progress resulting in a reduction in efficiency and productivity compared with the levels normally expected. See Section 5.8
DMS	The abbreviation for 'document management system'
document management system	A 'database' (q.v.) with automated input of document-specific values, often with a graphical interface which facilitates the viewing of a scan of the document
domestic subcontractor	A subcontractor working under the direction of and chosen by the contractor (q.v.)
double entry	A method of book-keeping in which the data are recorded in different places by duplicating the data input
down time	The period (usually brief) when work is suspended
driving relationship	A relationship by which either the logical start or logical finish of an activity is dependent upon the start or finish of another activity. See also 'non-driving relationship'
DTM	The abbreviation for 'dynamic time model'
dummy	A nominal task which requires no work to be carried out in an ADM (q.v.) network; can also be used in PDM (q.v.) to simplify logical connections between activities. See also Section 4.10.8.7
duration	The period between the start and finish of an activity
dynamic schedule	A schedule which will react to change and predict the consequences of that change (see 'dynamic time model')
dynamic time model	A critical path network of which the essential characteristics are:
	■ a high quality critical path network (q.v.) without any constraints (q.v.) that will inhibit the schedule from reacting dynamically to change

■ a combination of the short-term look-ahead using resource and location-based logic in High Density (q.v.) with the schedule for the whole of the work using activity-based logic in Medium Density (q.v.) and Low Density (q.v.)

■ activity durations in the High Density part of the network calculated by reference to the planned resources and their expected productivity

■ the High Density part of the network updated with records of the resources actually used and the amount of work actually achieved in the update period

■ the schedule being revised to incorporate fuller and better information as it become available on a 'rolling wave' (q.v.) principle going forward so that the next three months' work is always scheduled in High Density

■ the schedule being impacted by intervening events (q.v.) that have occurred, are occurring, or are likely to occur in the future, so that their predicted effect can be managed

■ see also Part 3 and Appendix 4 for a description of the dynamic time model in use

dynamic time modelling	The act of modeling the predictive time and cost effects of change on a critical path network using a dynamic time model (q.v.)
early date	The earliest that an activity is calculated to start or finish in a critical path network (q.v.)
earned schedule	An extension to earned value management theory and practice, that corrects known errors in the calculation of schedule variance and schedule performance index, in order to estimate (on the basis of an extrapolation of the productivity achieved to date) the likely key dates, sectional completion dates and the completion date
earned value analysis	A method of deducing a departure from planned progress by comparing the value of work done in relation to its planned value at a data date (q.v.)
earned value management	The process of managing a project by monitoring the value of work done in relation to its planned value at a data date (q.v.). See Section 5.9
earthworks	The term applied to describe the 'cut and fill '(q.v.) or regrading of land
effort expended	The amount of work done in relation to the whole
egress	The exit or way out
employer	The party which agrees to pay for the consultancy services and/or construction works. Under standard forms of construction contract, the employer may be variously referred to as 'the government',[2] 'the owner',[3] 'the client',[4] 'the authority',[5] 'the purchaser'[6] or 'the principal'[7]

---

[2] Contracts construed by the US Boards of Contract Appeals.

[3] American Institute of Architects standard forms.

[4] Association of Consulting Architects, Standard Form of Contract for Project Partnering, 2000, and later editions.

[5] The pre-1998 versions of the UK government standard forms of contract.

[6] JCT Constructing Excellence Contract, 2006, Institute of Chemical Engineers lump sum contract, 2001 and Institution of Engineering and Technology, Model Form 1, lump sum contract, 4th ed., 2000.

[7] Australian Standard Conditions of Contract 4th ed., 1992, AS4000 Australian Standard Conditions of Contract 199, Government of New South Wales, General Condition of Contract, and New South Wales Government GC21 (Edition 1) General Conditions of Contract July 2003.

employer's contractors	Those contractors engaged by the employer (q.v.) (usually for specialised work) under a separate contract from the main contract for the works
employer's goods and materials	Those goods and materials to be provided to the contractor by the employer (q.v.)
end-user requirements	The needs of those who are to use the finished product
engineer, procure and construct	Sometimes called 'turnkey'; a form of contract in which the contractor adopts the obligation of fitness for purpose in designing and providing the finished product ready for use
engineering logic	The sequence or order in which tasks must be carried out based on scientific laws and principles. See Section 4.23.4.1
environmental conditions	Usually weather conditions, but can be other conditions in which activities have to be carried out, e.g. mines and some large-scale, heavy-engineering projects (dams, nuclear power stations and the like)
EPC	The abbreviation for 'engineer, procure and construct'
ES	The abbreviation for 'earned schedule'
EVA	The abbreviation for 'earned value analysis'
event register	A listing of intervening events and the salient information concerning those events
EVM	The abbreviation for 'earned value management'
executive summary report	A management report summarising the key milestone and completion dates, main activities and deliverables of the (WBS level 1) Low Density schedule
EXF	The abbreviation for 'expected finish'
expected finish	See Section 4.27.3.11
expired time	The period elapsed in relation to the planned duration
extension of time	An award to the contractor (q.v.) of more time to complete where the contract permits the contract completion date to be deferred to a later date for delays that are not of the contractor's making. "The benefit of an extension of time is only to relieve the contractor of liability for damages for delay (usually liquidated damages (q.v.)) for any period prior to the extended contract completion date. The benefit of an extension of time for the employer (q.v.) is that it establishes a new contract completion date, and prevents time for completion of the works being at large" [8]
FF	The abbreviation for 'finish-to-finish'
field	A category of data in a database (q.v.) in which values can be held for filtering (q.v.) and sorting (q.v.)
filter	A facility allowing subsequent commands to operate only on records that conform to specified criteria[9] in an electronic database for identifying the values in one or more fields amongst others
finish-no-earlier-than	See Sections 4.27.3.5 and 4.27.3.12
finish-no-later-than	See Sections 4.27.3.6 and 4.27.3.12
finish-to-finish	See Section 4.25.1.2
finish-to-start	See Section 4.25.1.3

---

[8] The Society of Construction Law Delay and Disruption Protocol 2002.
[9] Wideman Comparative Glossary of Common Project Management Terms v.3.1.

flag	A symbol intended to attract attention
flexible constraint	A logical constraint (q.v.) which can adopt either the earliest start and finish dates or the latest start finish dates. See Section 4.27.3
float	The degree of flexibility available for the scheduling of an activity; see also 'free float', 'total float', and 'negative float'. The amount by which an activity can be delayed without causing a delay to subsequent activities or the completion date. See Sections 4.28
float values	The quantity of float, expressed in time units available to an activity, milestone or event
floor slab	A dense construction forming a floor, usually of concrete
FNET	The abbreviation for 'finish-no-earlier-than'
FNLT	The abbreviation for 'finish-no-later-than'
force majeure	An intervening event (q.v.) which is outside the control of both the contractor (q.v.) and the employer (q.v.) and for which, unless the contract provides otherwise, the contract is deemed frustrated
formulaic calculations	Calculations produced by formula
formwork	The temporary works required to constrain the pouring of concrete
fragnet	A subnetwork of activities modelling an event or a section of the works; see Section 5.8.36 and Appendix 8
free float	The period by which an activity may be delayed without delaying the start of any of its successor activities. See Section 4.28.4.1
FS	The abbreviation for 'finish-to-start'
gang	See 'crew'
Gantt chart	See 'bar chart'
generic resource	Resource defined by trade or undertaking, e.g. bricklaying
get milestone	A milestone (q.v.) that represents the commencement of a party's work in relation to the work carried out by a predecessor party
give milestone	A milestone (q.v.) that represents the completion of a deliverable, sufficient to permit works to be carried out by another party to commence
GMP	The abbreviation for 'guaranteed maximum price'
ground beam	A beam linking foundation pads together
guaranteed maximum price	A form of target-cost procurement in which it is said that more risk is borne by the contractor than in other forms of contract
hammock	An activity bar which hangs between the start of the earliest and the finish of the latest in a group of activities, often used for summarisation purposes
hard logic	See 'engineering logic'.
health and safety legislation	The laws of a State governing the health and safety of site operatives and others in the workplace
health and safety manager	The party responsible for managing health and safety matters and ensuring that those responsible comply with health and safety legislation
High Density	That part of a critical path network in which the activity durations are calculated using resource and productivity data and the activities are networked using resource- and location-based logic.

horizontal traceability	A state in which the schedule is rational, has been planned in a logical sequence, accounts for the interdependence of detailed activities and planning packages, and provides a way to evaluate current status. Schedules that are horizontally traceable depict logical relationships between different elements and their completion
ID	The abbreviation for 'identification data'
identification data	See 'activity identification code'
impacting	The process of calculating the effect of an intervening event (q.v.) on a dynamic schedule (q.v.); see Sections 5.8.34 to 5.8.43
implied variation	An act or omission which is deemed to be a variation
industry standards	See Appendix 7
inflexible constraint	A logical constraint (q.v.) which, if linked to an activity or resource in a CPM (q.v.) schedule, will prevent it from reacting according to the logic (q.v.) of the schedule. See Section 4.27.3.9
information flow	The transfer of information from one party to another
information release dates	Dates upon which the design team (q.v.) are bound to release information to the contractor, under the contract
information release schedule	A document (required under some forms of contract) setting out the dates on which the employer warrants that information from the design team will be released to the contractor
information request schedule	A document setting out the information requested by the contractor and the date by which it is required to be provided
interface schedule	A schedule that collects 'give' and 'get' milestones (q.v.) from a number of schedules and coordinates them in a coded database
interruptible activity	An activity in a Low Density (q.v.) or Medium Density (q.v.) part of a schedule which, in order to conform to the logic of the network, requires a duration longer than its contiguous duration (q.v.) or which, depending upon the intermittent availability of resources, can be stretched or made longer than the original estimate
intervening event	An event which interferes with the progress of the work; see Appendix 1 for a schedule of those usually at the risk of the employer (q.v.); see also Section 5.8.34.43
IRS	The abbreviation for 'information release schedule' (q.v.), but see also 'information request schedule'
jagged line	The term used to describe a comparison of planned and actual progress on a schedule, by drawing a vertical line at a given date that is jagged where progress has deviated from that planned. See Section 5.9.21
key date	A term usually used for the date upon which a work stage is to be completed or the works are to have achieved a defined status (e.g. 'watertight')
key performance indicator	A measurable value used to evaluate whether a target is achieved
KPI	The abbreviation for 'key performance indicator'
labour	The human resource
ladders	A sequence of overlapping activities connected with both a start-to-start and a finish-to-finish link (or their equivalent in an ADM network (q.v.)) at their starts, finishes, or both. See Sections 4.26.3.6 and 4.30.7.33

lag	A number of time units applied to a link that affect the schedule calculations and may impose a delay on the successor activity. The amount of time after one activity is started or finished before the next activity can be started or finished. See Section 4.26
lagged finish-to-finish	See Section 4.26.3.1
lagged finish-to-start	See Section 4.26.3.2
lagged start-to-start	See Section 4.26.3.3
late date	The latest that an activity is calculated to start or finish in a critical path network (q.v.)
lead	See 'negative lag'
learning curves	The line described by the cumulative graph of productivity in a repetitive cycle as the workforce become more proficient
level crossing	A junction between one method of transport and another (usually a railway and a road) on the same plane, with the result that in order to avoid collisions, when one is in use the other must be barred from use
levelling	See 'resource critical resource levelling'
limited possession	A short period in which the contractor is entitled to sole access to a particular part of the work
line-of-balance diagram	An illustration of sequence of repeated tasks across a number of work areas identifying resource, location and time used to optimise the use of resources across all trades working in defined areas (q.v.). See Section 4.9.8.2
linked bar chart	A CPM network illustrated as a bar chart which shows dependency links between the activities. See Section 4.10.8.12
liquidated damages	A sum of money agreed as a rate per calendar period that represents, at the date of contract, no more than the loss which the employer (q.v.) is reasonably likely to suffer in the event that, as a result of the contractor's default, the works (q.v.) are not completed by the contract completion date
local regulations	Laws of authorities in the locality of the works
lockout	Exclusion of the workforce from the works by the employer
logic	The principles underlying the arrangements of activities and milestones in a critical path network by which every activity has a predecessor (except for the commencement milestone) and a successor (except for the completion milestone) so as to identify one or more critical paths
logic tracing	The process of tracking a path by reference to driving relationships (q.v.)
logical interface	A predecessor or successor
logistics	Management of the flow of resources from procurement to completion of the works
long lag	See Section 4.30.7.26
longest path	See 'critical path'
Low Density	That part of a critical path network containing activities for which all information necessary for their execution is not available. At Low Density, activities are networked using activity-based logic
machines	See 'plant'
mandatory-finish	See Section 4.27.3.11

mandatory-start	See Section 4.27.3.11
manually applied constraint	A constraint applied to a network arising other than as a result of the logic (q.v.); see also Section 4.30.7.18
master schedule	A schedule that combines and synchronises with other schedules created as subprojects. See Section 4.8
Medium Density	That part of a critical path network containing activities for which all information necessary for their execution is available, other than, where relevant, the required resources and/or their estimated productivity. At Medium Density, activities are networked using activity-based logic
method statement	See 'planning method statement' and/or 'construction method statement'
MF	The abbreviation for 'mandatory-finish'
milestone	A point in time identified on a schedule which has neither duration nor resources attributed to it
milestone chart	A schedule containing only milestones, in a single chart
milestone management	See Section 4.8.4.7
milestone monitoring	See Section 5.9.28
milestone schedule	See Section 4.8.4.7
mitigation	Action taken to alleviate predictable loss, expense or delay
mobilisation period	The lapsed time needed to assemble resources between instructions issued to perform an activity and its commencement
moderate constraints	See Section 4.27.3.3
monitoring	See 'progress monitoring'
Monte Carlo analysis	A method for calculating the probabilities of outcomes by simulation, running a model many times, using a computer.[10] See Section 4.9.9.2
MS	The abbreviation for 'mandatory-start'
must-finish-on	See 'mandatory-finish'
must-start-on	See 'mandatory-start'
negative float	The amount of time by which an early date (q.v.) of an activity exceeds its latest date (q.v.).[11] See Section 4.28.4.3
negative lag	The relationship between a predecessor and its successor activity in which the successor is required to start a period of time before the predecessor. See Sections 4.26.3.8 and 4.30.7.30
network	A schedule in which the activities are linked by their logical predecessors and successors
network diagram	See Section 4.10
network logic	The collection of activity dependencies that make up a network diagram (q.v.)
node	The start and end of activities in an activity on arrow network (q.v.) or the activity box in a precedence network (q.v.). See also Sections 4.10.8.6 to 4.10.8.9

---

[10] Wideman Comparative Glossary of Common Project Management Terms v.3.1.
[11] Ibid.

nominated subcontractor	A subcontractor selected by the employer (q.v.) to work under the direction of the contractor (q.v.) usually for work of a specialised nature
non-contiguous activity	See 'interruptible activity'
non-driving relationship	Logic (q.v.) required to close fully define a network (q.v.) and ensure the absence of open ends (q.v.) to an activity which is not on the critical path (q.v.) and where the presence of the link does not affect the scheduling of its successor. See also 'driving relationship'
OBS	The abbreviation for 'organisational breakdown structure'
occupational commissioning schedule	The employer's schedule showing how it will occupy and use the building or facility, containing details of possessions, furnishing, commissioning and testing, access and training. See Section 4.6.2.7
open end	See Section 4.23.1
organisation breakdown structure	An organisation chart depicting the personnel involved in the project, their role and seniority. See also Section 4.11.10
organising	Putting into the required order or sequence
out-turn cost	The ultimate cost of a project to the employer. The tender cost plus the cost of variations and compensation for loss and/or expense, including consultants' fees, planning fees and licenses
out-turn time	The ultimate duration of a project from inception to completion, including the time taken up by delay, howsoever caused
overload	The status of a project in which, in order to meet a duration, resources are required in excess of that planned
overtime	Time required to be worked in excess of the regular or normal hours of work
owner	See 'employer'
pacing	Slowing the progress of work by the reduction of resources or suspension of work in order to keep pace with delayed work
package contractor	See 'trade package contractor'
parametric object	A building element within a BIM model that is synchronised with another to which it is related, so that if one changes the other will also change
partial possession	Deemed or actual possession by the employer (q.v.) of part of the works (q.v.), or section of the works
PDF	The abbreviation for 'portable document format'
PDM	The abbreviation for 'precedence diagram method'
phase	A major period of work in a project concluding in a milestone (q.v.)
planned value	The estimated worth of the work to the party paying for it
planning	See 'project planning'
planning method statement	A report prepared by the party responsible for planning the work, describing the strategies to be employed in the project delivery and giving a description of the assumptions and calculations underpinning a schedule and the changes made to it. See Section 2.1
planning strategy	See Section 2.4
plant	Mechanical equipment

portable document format	A file format used to present documents in a manner that is independent of application software, hardware, and operating systems
possession	Transfer of the normal operations and workings of an employer's facility or part of a facility into the control of the contractor to enable safe working (e.g. highway lane or railway). Control of an area (q.v.) or zone (q.v.) of operation
precedence diagram	A network in which the activity is on the node (q.v.)
predecessor	The activity or milestone that controls the start of a link in a PDM network (q.v.). That part of an activity which must be completed before another activity can start or finish. See also 'successor'
predicted cost	The estimated out-turn cost to the party paying for the work
preferential logic	See Section 4.23.4.2
prime cost sum	A contingent sum of money included in the contract sum for work to be carried out by others under the contractor's (q.v.) direction or for materials yet to be specified
production rate	The quantity of work achieved in a unit of time. Standard rates are contained in published pricing or estimating literature or from an organisation's own historical databanks
production records	See 'progress records'
productivity	A measure of the efficiency of production
productivity analysis	The study of the rate of efficiency achieved or amount produced over a period of time
productivity data	A quantified measure of work output per unit of labour or plant in unit time
productivity quotient	The rate at which a quantity of work can be accomplished with a given resource in relation to a specific activity
programme	The time-control document required by some forms of contract, usually in printed form; see also 'schedule'
programming	The act of producing a programme (q.v.) see also 'scheduling'
progress	The progression of development showing the relationship between current conditions and past conditions[12]
progress monitoring	The process of reviewing actual performance, over a period of time, against the planned progress over the same period
progress record	The accepted database record of the resources actually used to perform a recorded quantity of work. See Section 5.9
progress override	A facility provided by some software products in which out-of-sequence progress is given priority over the logic of the network, when recalculating the critical path. It produces unreliable results (see Section 4.30.7.39)
progress records	The accepted database records of the resources actually used to perform a recorded quantity of work. See Sections 5.1, 5.5 and 6.6
progress update	The addition of progress records to a critical path network to establish the effect of progress achieved on the timing and sequence of activities planned to be carried out in the future. See Sections 2.5 and 5.6
project	A temporary organisation established to accomplish an objective

---

[12] Wideman Comparative Glossary of Common Project Management Terms v.3.1.

project budget	The amount of authorised capital allocated to a project and its distribution to elements, based on an estimate of cost plus profit calculated by reference to the design arising out of the project brief
project management	The application of knowledge, skills, tools and techniques to lead and direct the project
project manager	The person (or people) appointed to lead and direct the project (q.v.) on behalf of its stakeholders
project manager's report	A periodic report usually issued monthly advising senior management and the employer of the status of the project
project planning	The process by which the design team, construction management team and the project planner create the delivery strategy for the project. See Sections 1.1.5 and 2.4
project scheduler	See Section 1.7
project scope	All the work to be carried out in order to achieve the project's objectives
prolongation cost	The contractor's (q.v) time-related site establishment or overhead costs during a period of delay to its planned completion date
provisional sum	A contingent sum of money included in the contract sum for work of which the detail cannot be fully described at the time of tender
rebar	See 'reinforcement'
record-keeping	See Sections 2.5 and 5.4
record retrieval	Access for use of stored information
recovery	Making up of lost time at the contractor's expense; see also 'acceleration'
reinforcement	Steel bars bedded in concrete to add tensile strength; also known as 'rebar'
remaining time	The duration planned to elapse before an activity is completed
repetitive cycle	A sequence which is carried out more than twice
resequencing	A change to the order and timing of future activities. This can occur automatically as a result of progress being different from that planned, where the scheduling software will recalculate the start and finish dates of the remaining activities based on the historical progress achieved
resource	Anything necessary for the achievement of work. but typically materials, labour, plant, space, cost
resource aggregation	Summation of the requirements for each resource, and for each time period[13]
resource allocation	Information about the use of resources by an activity
resource calendar	A calendar that defines the working and non-working patterns for specific resources[14]
resource critical resource levelling	A method of strategic resource allocation in which potential resource overloading is reduced by showing a change in the project end date to reflect the time required to complete the project with the available resources
resource loading	The process of adding resource data to the activities within a schedule

---

[13] Wideman Comparative Glossary of Common Project Management Terms v.3.1.
[14] Ibid.

resource logic	The reason for resources to proceed in a particular sequence. See also preferential logic. See Section 4.23.4.3
resource planning	The selection of resources in type and quantity to complete a project[15]
resource scheduling	The process of calculating a set of scheduled dates that takes into account the resource requirements of each activity and the availability of these resources.[16] See Section 4.20
retained logic	A facility provided by some software products in which the logic of the network is given priority over out-of-sequence progress, when recalculating the critical path. It produces unreliable results. (See Section 5.6)
rework	Repair of defective work
right of way	An easement granted or reserved over land for transportation purposes. Typically used for the purpose of giving the contractor access to construct the works in connection with linear projects such as highways, rail transport, electrical transmission lines, oil and gas pipelines, and so on
risk management	The process of managing risk, including risk identification, risk analysis, risk reduction, risk transfer, risk avoidance and contingency planning
risk manager	The title given to the party responsible for managing the risk register
risk register	Schedule of foreseeable risks, likelihood of occurrence, possible consequences and planned remedial action
rolling wave	The process of scheduling in greater detail in waves as the project proceeds, available information increases, and the risks, assumptions, and milestones originally identified become more defined and reliable. See Part 3 for a description of the process
ROW	The abbreviation for 'right of way'
scaffolding	A temporary or movable platform or structure, usually constructed of tubular steel poles and timber boards, used to permit labour to work at high levels
schedule	See 'critical path network'
schedule density	See Section 4.2
schedule integrity	The character necessary for the schedule to perform dynamically so as to properly calculate the consequences of change; see also Section 4.30.7.12
schedule logic	See Section 4.23
schedule performance index	See Section 5.9.42
schedule review	See Section 5.7
schedule variance	The difference between the current schedule and that used for comparative purposes
scheduler	See Section 1.9
scheduling options	Software switches which select different computational algorithms (q.v.). See also Section 4.30.7.37
s-curve	A graphic display of cumulative costs expended (labour hours or other quantities) plotted against a time line of their expenditure. The name derives from the S-like shape of the curve, which tends to be flatter at the beginning and end, but steeper in the middle

---

[15] Ibid.
[16] Ibid.

sectional completion date	The date by which a defined part of the works is contractually bound to be completed
section manager's report	A Level 4 report showing progress against the detailed level working schedules
separate contractor	A contractor employed to carry out work under a different contract from that for the main works
SF	The abbreviation for 'start-to-finish'
short-term look-ahead	A programme (q.v.) or schedule (q.v.) showing the activities to be under-taken in the near future
short-term look-ahead report	A narrative describing the High Density part of the schedule from the data date (q.v.) going forward
shuttering	See 'formwork'
slack	See 'float'
smoothing	See 'deadline critical resource levelling'
SNET	The abbreviation for 'start-no-earlier-than'
SNLT	The abbreviation for 'start-no-later-than'
soft logic	See 'preferential logic'
sorting	The process of defining the order in which values will be displayed in a database report
SPI	The abbreviation for 'schedule performance index'
spreadsheet	An electronic grid of cells in which alphanumeric data or formulae can be located and interrelated for listing, filtering, sorting, organisation and calculation
SS	The abbreviation for 'start-to-start'
staging	See 'scaffolding'
standard outputs	Published productivity quotients (q.v.). See Appendix 7
start-no-earlier-than	See Section 4.27.3.7
start-no-later-than	See Section 4.27.3.8
start-to-finish	See Section 4.25.1.4
start-to-start	See Section 4.25.1.1
statutory approvals	Permissions given by a body authorised by law to issue such permissions
statutory undertaker	A company authorised by law to provide services usually concerned with transport or the provision of water, gas, electricity and so on. See also 'utilities'
stretched activity	See 'interruptible activity'
strike	A group action for withdrawing labour from the work
strike formwork	The dismantling and removal of formwork (q.v.)
subcontractor	See 'domestic subcontractor' and 'nominated subcontractor'
submission	An application for approval or consent
subproject	A part of the works with a start and completion date

successor	The activity or milestone that is controlled by the completion of a link in a PDM network (q.v.). See also 'predecessor'. That part of an activity which logically cannot start or finish until after another activity has started or finished
supply chain	The collection of organisations providing goods and services in the manufacture of raw materials through to the completed product and their integration into the works (q.v.)
SV	The abbreviation for 'schedule variance'
target schedule	A schedule against which current progress is measured and managed
task	An individual item of work that can be an activity or part of an activity (q.v.)
temporary works	Work which must be carried out in order to construct the permanent works, but which is not intended to remain
tender	An offer to carry out work for compensation under a contract; also known as a 'bid'
tender schedule	The contractor's first schedule produced in connection with its tender. See Section 4.6.2.2
testing and commissioning	The process of validating and adjusting the permanent works, or any part of it, and rendering it fit for use
TF	The abbreviation for 'total float'
third party	Any party other than the two parties to the contract
third-party issues	Those matters requiring action relating to work to be carried out by other contractors or companies in connection with the works
third-party projects	Work required to be carried out, other than that which is the subject of the contract between the contractor and employer (q.v.)
tie-in	A connection between different elements, typically, for example, between a new part of a road or railway with the existing road or railway
time chainage diagram	A diagram with one axis representing time and the other axis distance. See Section 4.10.8.3
time contingency	A period of time allowed for work or suspension of work which may or may not be required
time contingency buffer	See 'time contingency'
time impact analysis	A method of identifying the cause of delay to a key date, sectional completion date or completion date by sequentially isolating the effect of delay to progress caused by slippage at the contractor's risk from that caused by intervening events at the employer's risk, in chronological order of their initiation using a critical path network
time management	See Sections 1.1 and 1.10
time model	See 'dynamic time model'
time now	See 'data date'
topping-out	The term given to the completion of the last part of a structural enclosure
total float	The amount of time an activity can be delayed without delaying a key date, sectional completion date or the completion date, without contravening a constraint. See Section 4.28

trade	A particular specialised type of work
trade package contractors	Contractors engaged by the employer to carry out a defined part of the works under a construction management contract (q.v.)
trailing open end	An activity without a successor to its finish; see also Section 4.30.7.22
triangular distribution	See Section 4.9.9.4
turnkey	See 'engineer, procure and construct'
unexpired	A time allowance which is unused; see also 'contingency'
utilities	The public-company suppliers of water, gas, electricity, communications and other publicly available services. See also 'statutory undertaker'
variation	An instructed change as defined under the contract
vertical traceability	The consistency of dates, status, and scope requirements between different levels of a schedule at Low Density, Medium Density and High Density. When schedules are vertically traceable, higher density parts of the schedule are clearly consistent with lower density milestones, allowing for total schedule integrity and enabling different teams to work to the same schedule expectations
WBS	The abbreviation for 'work breakdown structure'
windows method	The expression used to describe a critical path (q.v.) method of delay analysis by reference to predetermined periods of time. See also Section 5.8.39
work breakdown structure	A hierarchal breakdown of the total scope of work, defining the work packages and activities at a level above that illustrated by the network (q.v.). See Section 4.11
working schedule	The contractor's schedule for the works. See Section 1.1.6
work pattern	The sequence of working periods and non-working periods in a working day
works	The work the contractor is required to carry out under the construction contract
works package contractor	See 'trade package contractor'
work type definition	A description of work in sufficient detail to render it unique amongst other work descriptions of like type
ZFF	The abbreviation for 'zero-free-float'
ZTF	The abbreviation for 'zero-total-float'
zero-free-float	See Section 4.27.3.4
zero-total-float	A logical condition in which the early and late dates are set the same so that the activity is shown to be critical. See Section 4.27.3.12
zonal logic	A form of preferential logic in which one part of the physical work is required to precede another. See Section 4.23.4.4
zone of operation	A division of the work for the purposes of management and control

Index

Acceleration
    by adjusting contingencies   2, 11, 16, 19, 21, 29,
        31, 32, 37, 53, 54, 109, 139, 146
    by change of design   138
    by change of information flow   139
    by change of method of working   138
    by change of resources   31, 139
    by change of sequence   2, 11, 16, 19, 21, 29, 31,
        32, 37, 53, 54, 109, 138, 139, 162
    by change of work content   138
    by change of working time   139
    cost of   38
    definition of   191
    monitoring of   102
    by omission of work   31, 138
    proposals for   102
    timing of   21, 139
    see also Recovery
Activity
    content code (see Activity content code)
    cost code (see Activity cost code)
    definition of   191
    description (see Activity description)
    dummy   59, 196
    duration (see Activity duration)
    identifying data (see Activity ID)
    logic (see Activity logic)
    open ended   102, 103
Activity content code
    adding and amending   76
    definition of   75, 191
    design considerations for   45, 48, 71, 75, 76, 124
    filtering and organisation of   75, 76
    group coding   76
    integrity of   101
    in multiple tasks   75
    in procurement   75
    software issues   75, 76, 182
    use of   75, 76, 124
Activity cost code
    audit of   101,
    definition of   76, 191
    design considerations   45, 48, 77

    filtering and organisation of   77
    group coding   77
    at High Density (see High Density)
    integrity of   101
    at Low Density (see Low Density)
    at Medium Density (see Medium Density)
    software issues   75, 76, 182
    use of   76–77
Activity description
    amending   122
    design considerations   71–75, 115, 122
    at High Density (see High Density)
    at Low Density (see Low Density)
    at Medium Density (see Medium Density)
    software issues   74
    uses of   59, 61, 74, 115, 122
Activity duration
    audit of   101, 102
    calculated from resources, xiii   5, 26, 67, 78, 79,
        80, 81, 82, 85, 116, 121, 175
    definition of   191
    estimated from benchmark (see Benchmarking)
    estimated from comparison   78, 79, 81
    estimated from experience   78, 79, 81
    estimated from formula   54, 55, 56
    estimated from historical data   78, 81
    estimated from industry standards   78, 79
    estimated from specification   78, 87
    estimated from unspecified data   5, 26, 78
    at High Density (see High Density)
    intermittent   78, 104, 105
    at Low Density (see Low Density)
    at Medium Density (see Medium Density)
    software issues   78, 104, 178, 180
Activity ID
    coding structure   70–73
    definition of   70, 191
    design considerations   70–73
    at High Density (see High Density)
    at Low Density (see Low Density)
    at Medium Density (see Medium Density)
    software issues   71, 117, 177, 185
    uses of   24, 45, 61, 70–71, 88, 111, 115, 124

Guide to Good Practice in the Management of Time in Major Projects: Dynamic Time Modelling,
Second Edition. The Chartered Institute of Building.
© 2018 John Wiley & Sons Ltd. Published 2018 by John Wiley & Sons Ltd.